The FFT

Fundamentals and Concepts

The FFT
Fundamentals and Concepts

Robert W. Ramirez

Tektronix, Inc.

PRENTICE-HALL, INC., Englewood Cliffs, N.J. 07632

Library of Congress Cataloging in Publication Data

RAMIREZ, ROBERT W.   (date)
   The FFT, fundamentals and concepts.

   Bibliography: p.
   Includes index.
   1. Fourier analysis.   2. Fourier analysis—Data
processing.   3. Fourier transformations.   4. Fourier
transformations—Data processing.   I. Title.
QA403.5.R36 1985   515'.2433   84-8284
ISBN   0-13-314386-4

Editorial/production supervision and
    interior design: *Sylvia Schmokel*
Cover design: *20/20 Services, Inc.*
Manufacturing buyer: *Gordon Osbourne*

Printed in the United States of America

10 9 8 7 6 5 4 3 2 1

ISBN 0-13-314386-4 01

PRENTICE-HALL INTERNATIONAL, INC., *London*
PRENTICE-HALL OF AUSTRALIA PTY. LIMITED, *Sydney*
EDITORA PRENTICE-HALL DO BRASIL, LTDA., *Rio de Janeiro*
PRENTICE-HALL CANADA INC., *Toronto*
PRENTICE-HALL OF INDIA PRIVATE LIMITED, *New Delhi*
PRENTICE-HALL OF JAPAN, INC., *Tokyo*
PRENTICE-HALL OF SOUTHEAST ASIA PTE. LTD., *Singapore*
WHITEHALL BOOKS LIMITED, *Wellington, New Zealand*

Contents

Contents

Preface

Fourier analysis is not a new subject. It has been around since the early 1800s when J.B.J. Fourier developed the initial concepts and theory. Since then numerous papers and books dealing with Fourier theory have been published, and the Fourier series and integral have found their way into various college curricula.

So why another book on Fourier theory? Fourier analysis exists in a different context today. It used to be a pencil-and-paper issue, an interesting mathematical approach to getting frequency-domain information, but generally too difficult to apply in most practical cases. And even with the arrival of the digital computer, useful Fourier analyses were too time-consuming and computer-expensive for widespread use. Then, in the 1960s, J.W. Cooley and J.W. Tukey published *An Algorithm for the Machine Calculation of Complex Fourier Series*. Their algorithm became known as the *fast Fourier transform*, or *FFT*, and has become the new context for Fourier analysis. This is not just a digital context either, but a new context that allows quick, economical application of Fourier techniques to a wide variety of measurement and analysis tasks.

Thus the FFT is becoming a general analysis tool. FFT routines are found in most comprehensive software libraries, and FFT analyzers are becoming a more frequently encountered item. But even more than that, the FFT has joined forces with general-purpose instrumentation.

Today, instrument manufacturers are offering a variety of waveform digitizers, the most common type being what is generally referred to as a *digitizing oscilloscope*. These waveform digitizers are usually operated in conjunction with a minicomputer or desk-top calculator and a software library that often includes an FFT algorithm. The result is that Fourier analysis, as well as convolution and correlation, has been taken out of the textbook and put on the engineering bench. Because of its usefulness and increasing availability, it is expected that the FFT will become a major and commonplace measurement tool.

There is still, however, a missing link in the chain of events leading to general-purpose use of the FFT. That missing link is general familiarity with Fourier theory. You don't have to know all the details of the equations and their derivations. But you do need to know the concepts that they embody. To successfully use the FFT as a measurement tool, you do need to know what to expect in the frequency domain and how digital techniques affect the frequency domain.

For the most part, these concepts can be demonstrated through simple diagrams and pictures and can be discussed in simple terms. That is the approach taken in the following pages. Part I introduces classical Fourier theory with a slant toward later discussions of digital implementations. Part II covers the digital approach to Fourier analysis and makes heavy use of a waveform digitizer and an FFT algorithm to provide specific examples. Every effort has been made throughout to illustrate fully each concept and to discuss each concept in easy-to-understand terms. Part III provides a brief look at an FFT algorithm.

Why another book on Fourier theory?—to bridge the gap between classroom theory and practical use, and to do it in a language that people of different backgrounds and technical levels can understand.

ROBERT W. RAMIREZ

Acknowledgments

Though many solitary hours are spent putting words to paper, one can never truly write a book alone. Accordingly, I extend my appreciation to Tektronix, Inc., for providing the creative atmosphere as well as access to the instruments and software necessary for developing much of the material in this book. Also, I would particularly like to thank Lyle Ochs of Tektronix for his many suggestions. And, finally, my deep appreciation to my wife, Barbara, for her patience and encouragement throughout this project.

Part I ———————————————————————————

INTRODUCTION
TO FOURIER ANALYSIS

There are two chapters in Part I. The first is but a few pages—just enough to start you thinking about time and frequency as two related concepts. Though the relationship may not be intuitively obvious, an important relationship does exist. This relationship becomes more obvious in the second chapter, where the Fourier transform is explored. A good grasp of the concepts covered in these first two chapters is necessary to understand the digital analysis techniques covered in Part II.

Chapter 1 ————————————————

Time and Frequency: Two Bases of Description

Time. This is one of the fundamental concerns of people. How often during the day do we look at a clock or check our watches? Since birth, our lives have been geared to time. There is a time to wake up, a time to eat, a time to work, a time to play, and a time to go to sleep. We measure each day of our lives in time and use it to order the events that concern and affect us.

Time is universal. All people recognize its passage. All people live by it. Time, in itself, is central to many philosophical questions: Does time flow by us, or do we advance through time? And the measurement of time is an established science (horology) with a long history.

As far back as 3500 B.C., people were erecting poles and towers to cast shadows, the length of the shadow being an indication of the time of day. By the eighth century B.C., the Egyptians had refined this shadow concept to a fairly accurate sundial. They also developed water clocks to substitute during the night and on cloudy days. Later, the Romans and Greeks refined these devices further. But it wasn't until the fourteenth century A.D., that anything resembling a modern timepiece was developed. Then, in 1582, Galileo observed the constancy of a pendulum, and Christian Huygens, in 1665, incorporated Galileo's observation into the first pendulum clock. Until the advent of electrically driven clocks, the pendulum clock was the most accurate timepiece available. Now, by the 1967 agreement of the International Conference on Weights and Measures, the atomic clock is the ultimate standard.

TIME HISTORIES NEED TIME BASES

Today, one second is equal to 9,192,631,770 transitions between two specified, hyperfine levels of the cesium 133 atom.

But why so much precision in measuring the passage of time? The answer: Science demands it. A great deal of scientific theory is couched in terms of time histories. Furthermore, experimental proof of these theories requires time-domain measurements, and the precision of these measurements depends upon our ability to measure time.

As an example, Galileo reportedly used his own pulse as a timepiece in making his original pendulum observations. Each complete swing of a large pendulum took so many heart beats. With no greater precision than that, it was a natural experimental conclusion to say that a pendulum always shows the same simple harmonic motion. But theory tells us that this is not the case. In fact, for large displacements, the time for a complete swing of a pendulum is greater than for small displacements (Fig. 1–1). Proving this experimentally, however, requires a more precise timepiece than what Galileo had access to.

An electronic oscillator is in many ways analogous to a mechanical pendulum. The output of a sine-wave oscillator has a time history that closely resembles the time history of a pendulum's angular displacement. Galileo's concept of counting pulse beats can also be applied to measure the time for a complete voltage swing in an oscillator. The modern version of this concept is used in frequency counters. However, as shown in Fig. 1–2, an electronic pulse is used as a time base instead of a human pulse.

There is a time base involved in all time-domain measurements. In the case of Fig. 1–2b, the time base is used directly to measure the period of the signal. In other types of measurements, the time base is used to generate a time axis for an amplitude history. The precisely controlled speed of the paper drive for a chart recorder is an example of this latter case. Another example is the oscilloscope, which uses a ramp voltage to drive an electron beam at a constant rate across the face of a CRT (cathode ray tube). In both examples, the event or activity being captured for observation drives the pen or CRT trace in a direction normal to the time-base drive. The result is a time history of amplitude variation, as shown in the CRT photo of Fig. 1–3.

At this point, it might be well to pause and examine the CRT photo of Fig. 1–3 in a little more detail. This examination may seem trivial at first. But then it is surprising how much insight can be gained by starting with the seemingly trivial aspects of a subject. It is also surprising—no, embarrassing—to think of the number of measurements that get tripped up by trivia. So let's get on with the examination, which can justly be described as "attention to detail."

Referring to the CRT photo in Fig. 1–3, it is conventional to assign time zero to the left side of the display. Then, according to the time-base setting, time proceeds to the right. In Fig. 1–3, the CRT readout in the upper right portion indicates the time increments for each major division of the display. Vertical amplitude scaling

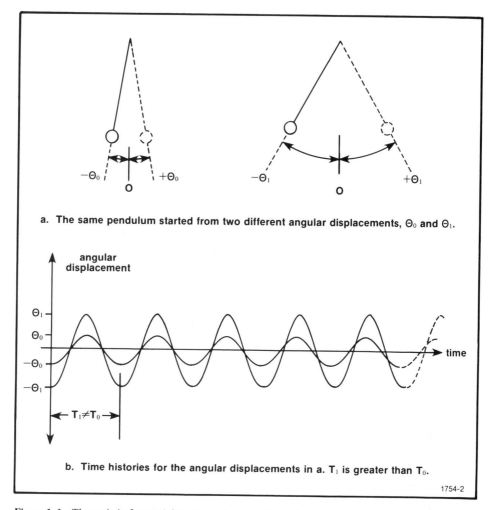

a. The same pendulum started from two different angular displacements, Θ_0 and Θ_1.

b. Time histories for the angular displacements in a. T_1 is greater than T_0.

1754-2

Figure 1–1 The period of a pendulum, T, varies according to its angular displacement, θ.

is indicated by the readout in the upper left corner. But what is the vertical amplitude referenced to?

In the case of Fig. 1–3, which is a photo of a waveform captured by a digitizing oscilloscope after a zero-referencing operation, the reference is indicated by the 0 DIV in the lower right corner. Here, the 0 DIV refers to the vertical center division as being the vertical zero reference. Other possibilities for zero reference might be above (for example, 3 DIV) or below (for example, −3 DIV) center.

With these three things defined—vertical and horizontal scale factors and a zero reference—the value of any point on the displayed waveform is defined. However, this is still not enough to fully define or describe a waveform.

To fully describe a waveform in time, it must not only be possible to pick off

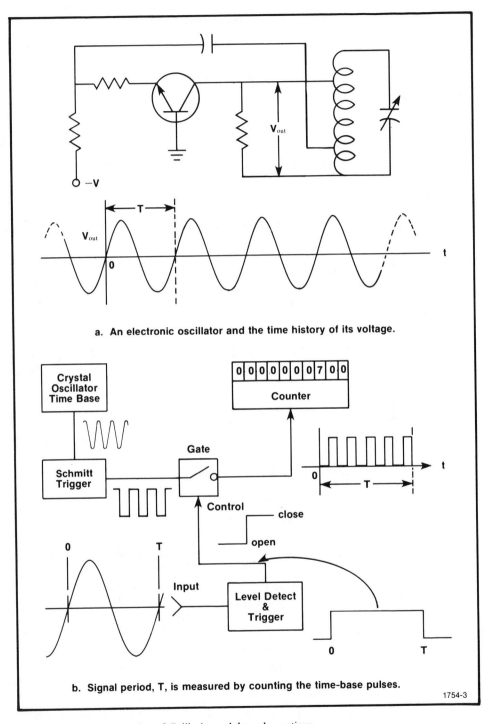

a. An electronic oscillator and the time history of its voltage.

b. Signal period, T, is measured by counting the time-base pulses.

1754-3

Figure 1–2 Electronic version of Galileo's pendulum observations.

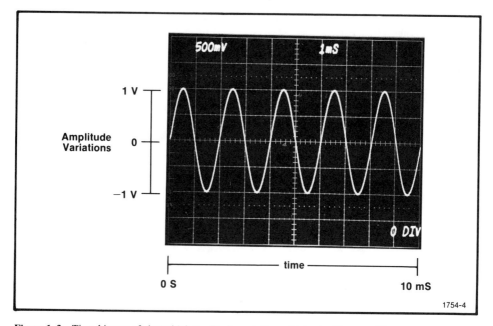

Figure 1–3   Time history of sinusoidal amplitude variations obtained with an oscilloscope.

its values at various points, but there must also be enough of the waveform displayed to discern its shape or type. For example, imagine the display if the sinusoid in Fig. 1–3 had been captured with a time-base setting of only 0.01 msec per division. That would stretch just the first tenth of a time division to cover the entire display. The waveform would appear to be a ramp instead of a sinusoid!

However, the CRT photo of Fig. 1–3 does give a complete time-domain description of a sinusoid. It is complete because it is fully scaled in time and at least one full repetition of the waveform is displayed. From this it can be assumed that the waveform is sinusoidal, at least within the bounds of the display area. What happens outside the display is not recorded and, therefore, is actually undefined. In the case of Fig. 1–3, however, experience and common sense lead us to assume continuation of the waveform in the same manner beyond the confines of the display. And so, we have a time history of a sinusoid.

SINUSOIDS LOOK DIFFERENT
FROM A FREQUENCY VIEWPOINT

Once a sinusoid is completely described with respect to time, you can construct a new description of it with respect to frequency. This is shown in Fig. 1–4.

Figure 1–4 depicts a three-dimensional waveform space with amplitude as one axis and time and frequency as the other two axes. The time and amplitude axes

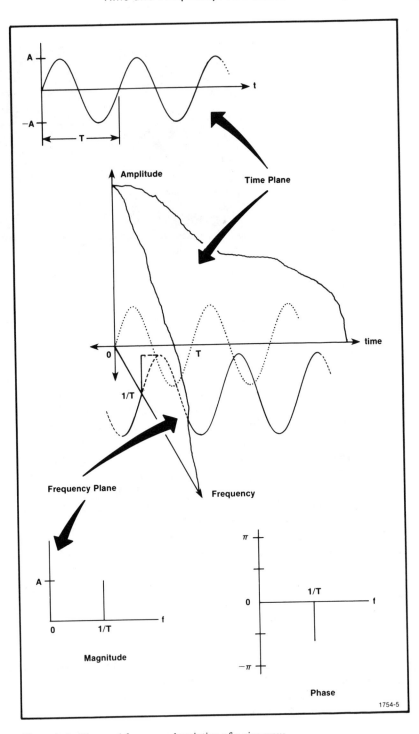

Figure 1–4   Time and frequency description of a sine wave.

define something that can be called a *time plane*. In the same manner, the frequency and amplitude axes define a *frequency plane* that is normal to the time plane.

The time history of a sinusoid, such as that in Fig. 1–3, can be treated as a projection on the time plane. In concept, the sinusoid can be thought of as actually existing at some distance from the time plane. This distance is measured along the frequency axis and is equal to the reciprocal of the waveform period.

Similarly, the sinusoid also has a projection onto the frequency plane. This projection takes the form of an impulse (a pulse of instantaneous rise and fall) with an amplitude equal to the sinusoid's amplitude. Because of symmetry, it is necessary to project only the peak amplitude rather than the full peak-to-peak swing. This is shown in Fig. 1–4 by the positive amplitude impulse on the magnitude diagram. The position of this impulse on the frequency axis coincides with the frequency of the sinusoid. (For now, just consider an impulse to be a line.)

The single impulse in the magnitude diagram defines both the amplitude and frequency of the sinusoid. With only this information, the sinusoid can be reconstructed in the time domain. Some additional information is needed, however, to fix the sinusoid's position relative to the zero time reference. This additional information is provided by a phase diagram, which also consists of an impulse located on a frequency axis. The amplitude of this latter impulse indicates the amount of phase associated with the sinusoid.

Phase diagrams for sinusoids can be determined by looking at the positive peak closest to time zero. For the case of Fig. 1–4, the positive peak occurs after time zero by an amount equal to one-fourth the period. There are 360° in a cycle or period, and the peak is shifted by one-fourth of this. So the phase in Fig. 1–4 is 360°/4, or 90°. Since the positive peak occurs after time zero, the sinusoid is said to be *delayed*. As a matter of convention, delay is denoted by negative phase. If the closest positive peak had been located before time zero, then the sinusoid would have been advanced. An advance is denoted by positive phase. The conventions are further illustrated in Fig. 1–5, and more examples are provided in Fig. 1–6.

In looking at Fig. 1–6, it should be pointed out that the total range of shift is −180° to +180°, or 360°. With no reference point fixed to the sinusoid, an actual shift out of the 360° = 2π range corresponds to a shift within the 2π range. For example, a sinusoid advanced by 360° + 90° = 450° is not generally distinguishable from the same sinusoid advanced by just 90°, so it can be represented as having just a 90° shift. This system of representing phase within a 2π range is referred to as *modulo 2π phase*. If on the other hand, a reference can be attached to the sinusoid, then shifts beyond the 2π range can be represented as such. This latter approach is referred to as a continuous phase representation and is detailed later in Part II.

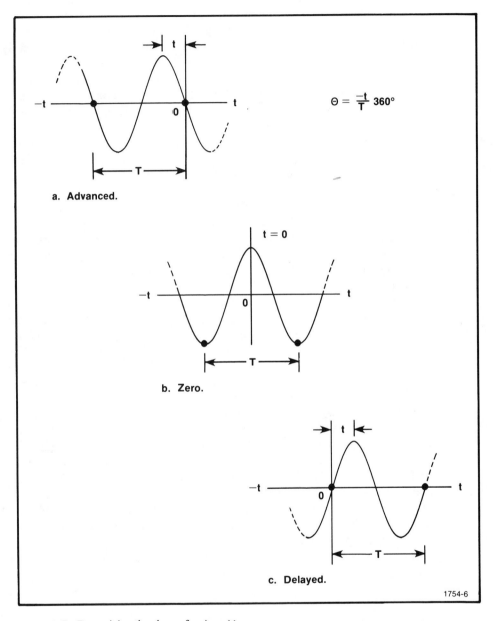

a. Advanced.

$$\Theta = \frac{-t}{T}\ 360°$$

b. Zero.

c. Delayed.

1754-6

Figure 1–5   Determining the phase of a sinusoid.

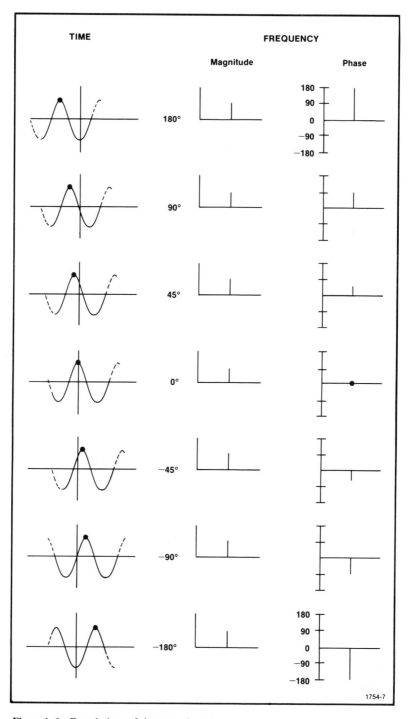

Figure 1–6 Descriptions of the same sinusoid at various phases.

NONSINUSOIDAL WAVEFORMS
ARE COMPOSED OF SINUSOIDS

By using the description conventions developed thus far, it is possible to build a variety of nonsinusoidal waveforms. For example, let's start with the frequency description of a sinusoid having a frequency of F_0 and a phase of $-90°$ (see Fig. 1–7a). Now let's take another sinusoid with a frequency of $2F_0$. Also, for the sake of illustration, let's say that the amplitude of this second sinusoid is half that of F_0. And, to add some interest, let's also give this second sinusoid a phase of $-45°$. This second

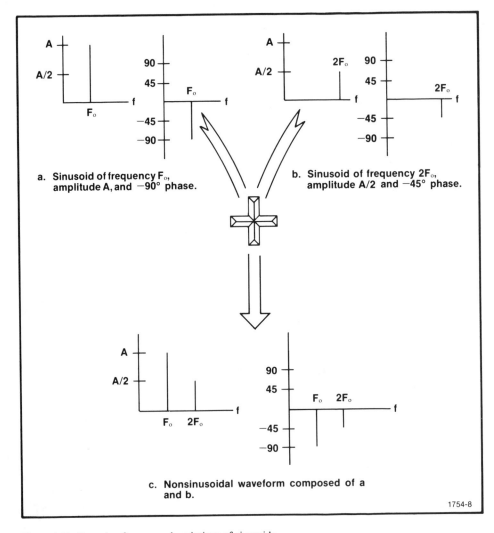

a. Sinusoid of frequency F_0, amplitude A, and $-90°$ phase.

b. Sinusoid of frequency $2F_0$, amplitude A/2 and $-45°$ phase.

c. Nonsinusoidal waveform composed of a and b.

1754-8

Figure 1–7   Summing frequency descriptions of sinusoids.

12

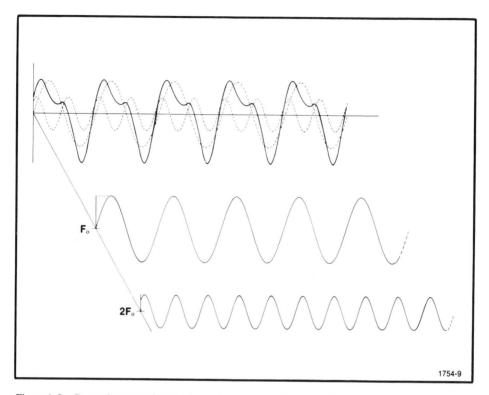

1754-9

Figure 1–8 Composing a nonsinusoidal waveform by summing sinusoids.

sinusoid is completely described in Fig. 1–7b, and the frequency description for the sum of the two sinusoids is shown in Fig. 1–7c.

To complete the picture, let's recast Fig. 1–7 in terms of three dimensions. This is shown in Fig. 1–8, where the concepts of projecting onto a time plane and a frequency plane are used. Additionally, the idea of summing multiple time-plane projections is introduced. Projections onto the time plane (or into the time domain, if you prefer) are shown by dotted lines, and their sum is indicated by a heavy solid line.

We can continue in the manner of Figs. 1–7 and 1–8, and by adding various sinusoids, various shaped projections onto the time plane can be obtained. Each shape, or waveform, is characterized by a unique combination of sinusoids. If any one of the sinusoids is changed in frequency, amplitude, or phase, the waveform's time-plane projection changes—it becomes a different waveform. This latter point can be illustrated by repeating the sum in Fig. 1–8 with the phase of $2F_0$ changed from $-45°$ to $+45°$. As can be seen in Fig. 1–9, this results in a different waveform.

For the most part, waveforms are measured in the time domain. Time-based measurements have historical precedence and are the most familiar data format. But time histories tell only one side of the story.

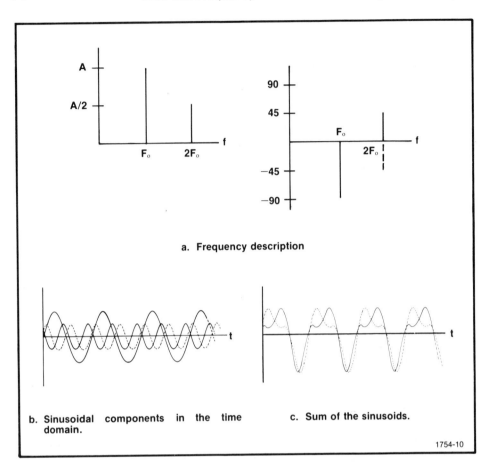

a. Frequency description

b. Sinusoidal components in the time domain.

c. Sum of the sinusoids.

1754-10

Figure 1–9 Changing any frequency component, in any way, changes the time-domain description: The phase at $2F_0$ is changed from $-45°$ to $+45°$. The original condition is shown by dotted lines and the new condition by solid lines.

Without a direct look at the frequency domain, waveshape changes are the only indication that some frequency components have been modified. For example, there's the rule of thumb that fast rise times imply high frequencies. So, if the square wave of Fig. 1–10a has some of its high-frequency components reduced or removed, we might expect to see a new waveform with a slower rise time. This is exactly what has happened in Fig. 1–10b, where a low-pass filter was used to attenuate the square wave's high-frequency content.

But how specific can you be about Fig. 1–10b? Some frequency axes are given in Fig. 1–10c. Can you decompose the waveform in Fig. 1–10b into its frequency components and indicate their magnitudes and phases on these axes? Can you indicate the amount of attenuation each component must have undergone?

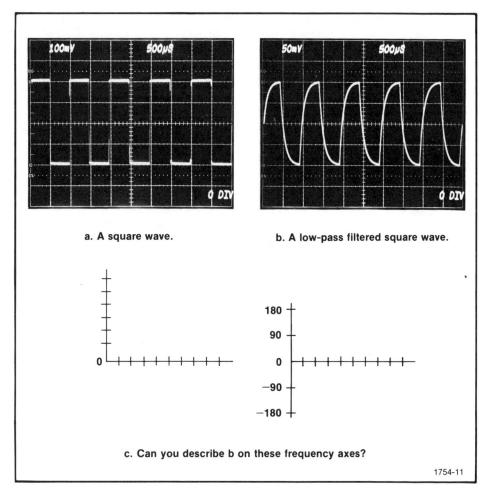

a. A square wave.

b. A low-pass filtered square wave.

c. Can you describe b on these frequency axes?

1754-11

Figure 1–10 The time domain does not tell the whole story.

GETTING INTO THE FREQUENCY DOMAIN

Something more than rules of thumb are needed to fill in Fig. 1–10c. If the waveform in Fig. 1–10b is available for measurement, it can be applied to a frequency-based oscilloscope. More often referred to as a *spectrum analyzer*, a frequency-based oscilloscope uses a filtering system to decompose a time-domain waveform into a magnitude diagram of its frequency components. A traditional spectrum analyzer does not, however, provide phase information. For many analyses, just knowing the frequencies and their magnitudes is sufficient. But, as was illustrated in Fig. 1–9, phase can be a crucial piece of information.

Another and more complete approach to getting frequency-domain information is to apply the rigorous mathematical technique known as *Fourier analysis*. This allows you to describe a time-domain waveform in terms of both frequency-domain magnitude and phase. Or, as will be seen later, Fourier analysis also offers the option of results in rectangular form, which consists of real and imaginary parts of the complex frequency domain. Unfortunately, however, the classic mathematical approach to Fourier analysis is frustrating for all but the simplest waveforms. If a waveform cannot be mathematically formulated, that is, expressed as an equation, classic Fourier techniques cannot be applied. A waveform can, however, be sampled and digitized with a waveform digitizer. When this is done, the discrete Fourier transform (DFT) can be used to Fourier transform the digitized waveform.

Since its general introduction in 1965, the fast Fourier transform (FFT) has been the commonly used algorithm for evaluating the DFT. This algorithm can be implemented through computer software, firmware, or hardware. Its major advantage is the speed with which it analyzes large numbers of waveform samples. And, combined with standard measurement concepts, the FFT effectively converts a time-based instrument to a frequency-based instrument.

Parts II and III of this book are devoted to digital techniques of Fourier analysis with the FFT algorithm. But before this subject can be approached, the continuous Fourier theory that it emulates must be understood. This Fourier theory for continuous, unsampled waveforms is the subject of the next two chapters.

Chapter 2 ─────────────────

Describing the Frequency Domain— The Fourier Series

Before asking what the Fourier series is, it might be well to ask how it came about. Why was this mathematical tool developed? Where did it come from?

The answers to these questions are not only an interesting piece of history, but they reveal a scientist of amazing capacity.

Born the son of an Auxerre tailor in 1768, Jean Baptiste Joseph Fourier grew to become one of France's major nineteenth-century administrators, historians, and mathematicians. His accomplishments began in 1798 when he went to Egypt with Napoleon. While there, he acted as an advisor on engineering and diplomatic matters and served as secretary of the *L'Institut d'Égypte* in Cairo. Also, he undertook an intensive study of Egyptian antiquities.

Fourier returned to France in 1801 and was appointed Prefect of the *Isére département* in 1802. He served at Grenoble in this capacity until 1814. During that time he was recognized as an able administrator, and Napoleon granted him the title of baron in 1809. This period also marked the beginning of Fourier's most important scientific contributions.

He contributed heavily to *Description de l'Égypte*, which covered the cultural and scientific results of Napoleon's invasion of Egypt. This work, issued in 21 volumes over the period from 1808 to 1825, contained much of the information Fourier had gained from studying Egyptian antiquities. The attention these volumes drew to the ancient Egyptian civilization resulted in Egyptology becoming recognized as a new and separate discipline.

Between his administrative functions and his contributions to Egyptology, it is amazing that Fourier still found time to do pioneering work in mathematical physics. His interest in heat conduction led him to begin work in 1807 on *Théorie analytique de la chaleur* (English translation, 1878, *The Analytic Theory of Heat*). The initial work was completed in Paris and published in 1822. It shows how a mathematical series of sine and cosine terms can be used to analyze heat conduction in solid bodies. The series that Fourier proposed, and which bears his name, is of the form

$$y = \frac{1}{2a_0} + (a_1 \cos x + b_1 \sin x) + (a_2 \cos 2x + b_2 \sin 2x) + \ldots$$

This, the Fourier series, was probably the first systematic application of a trigonometric series to a problem solution. Fourier spent the rest of his life working on his concept and expanding it to include the Fourier integral before his death in 1830.

Both the Fourier series and the Fourier integral allow transformation of physically realizable time-domain waveforms to the frequency domain and vice versa. They are the mathematical tools for what is now referred to as *Fourier analysis*.

Today, application areas of the Fourier series and integral transcend the original heat conduction application. For example, a few of the many areas of study benefiting from Fourier analysis are linear systems, antennas, mechanical vibration, optics, biomedicine, and various random process and boundary value problems.

It is interesting that such a far-reaching technique did not gain acceptability during Fourier's time. Various infinite series had been used prior to Fourier, most notably by Leonhard Euler, who found them an acceptable analysis tool. Many mathematicians, however, distrusted the use of the series, and one influential mathematician wrote in 1828: "Divergent series are the invention of the Devil, and it is shameful to base on them any demonstration whatsoever." Although Fourier's series is generally convergent, its validity did not escape the questioning of that era. Later work by P. G. L. Dirichlet (1805–1859), Georg Friedrich, Bernhard Riemann (1826–1866), Henri Lebesque (1875–1941), and others finally resolved any doubts about the validity of the Fourier series and integral. For our part, it is probably wise and most certainly expeditious to accept the conclusions of these great mathematicians and simply go on to understand the fundamentals and concepts involved in Fourier analysis, beginning with the Fourier series.*

CONDITIONS FOR EXISTENCE

If a Fourier series can be written for a waveform, then the components of the series completely describe the frequency content of the waveform. But there are some conditions that do have to be met.

The first condition that must be met for constructing a Fourier series is that the waveform be periodic. Precisely speaking, if the waveform is represented by $x(t)$ and there is a constant time, T, that exists such that $x(t) = x(t + T)$ holds for all time, t, then $x(t)$ is periodic with a period of T. In short, the waveform must repeat itself in time before a Fourier series can exist for it. Familiar examples of qualifying waveforms include sine waves, cosine waves, square waves, and so forth.

* "Fourier, Jean-Baptiste-Joseph, Baron," *Encyclopaedia Britannica* (1974), vol. 7, p. 577.

The most important thing to remember here, and for later discussion, is that periodicity is for all time. That is, to be periodic, the waveform must begin at minus infinity and repeat itself out to plus infinity. But the condition of periodicity is rarely, if ever, met precisely in the physical world!

No oscillators or pendulums existed at time equal to minus infinity. And if they did exist, it is doubtful their operation would continue to plus infinity. However, for the sake of practicality, the rules can be bent a little. Periodicity, for the purpose of writing a Fourier series, can be defined over an observable interval. In other words, a square-wave generator can be considered to have a periodic output from the time it is turned on to the time it is turned off. The Fourier series written for that output, however, describes the square wave as though it started at minus infinity and continued to plus infinity.

Beyond periodicity, the remaining conditions for existence of a Fourier series are referred to as the Dirichlet conditions. Briefly, they require that

1. If the function has discontinuities, their number must be finite in any period.
2. The function must contain a finite number of maxima and minima during any period.
3. The function must be absolutely integrable in any period; that is,

$$\int_0^T |x(t)| dt < \infty$$

where $x(t)$ describes that function.

These conditions, along with that of periodicity, establish the existence or nonexistence of a Fourier series for any $x(t)$.

There are some functions for which a Fourier series does not exist. For all practical purposes, though, the Dirichlet conditions are met when a periodic function accurately describes a physical occurrence. We can be confident that any real, existing oscillator has a frequency spectrum associated with its output.

THE FOURIER SERIES YIELDS DISCRETE SPECTRA

A few moments examining the square wave in Fig. 2–1a should satisfy any questions about the square wave's periodicity. From Fig. 2–1b through d, it can also be seen that the Dirichlet conditions are met. Thus the Fourier series for this waveform exists.

With the question of existence out of the way, the Fourier series for a square wave can now be written. However, it is not within the scope of this book to fully explore the writing of Fourier series. That subject is amply covered in a variety of texts. So only a brief synopsis is given here for reference and as a matter of definition. The greater interest here is in laying a foundation for building up to the concepts of the discrete Fourier transform (DFT) and the fast Fourier transform (FFT).

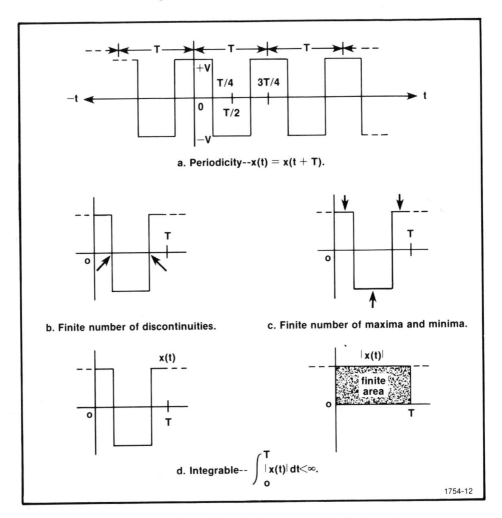

Figure 2–1   A square wave meets the conditions for the existence of its Fourier series.

The Fourier Series.   The general form of the Fourier series is $x(t) = a_0 + a_1 \cos \omega_0 t + a_2 \cos 2\omega_0 t + \ldots a_n \cos n\omega_0 t + \ldots + b_1 \sin \omega_0 t + b_2 \sin 2\omega_0 t + \ldots b_n \sin n\omega_0 t + \ldots$, where $\omega_0 = 2\pi f_0$. This is also frequently written as

$$x(t) = a_0 + \sum_{n=1}^{\infty} (a_n \cos n\omega_0 t + b_n \sin n\omega_0 t)$$

The Fourier series for a specific waveform is written by using salient features of the waveform to find specific values for the coefficients in the series above. First, ω_0 is found from the period of $x(t)$ and is equal to $2\pi/T$ (also, $f_0 = 1/T$). The a_0

coefficient is the DC (direct current) term and is equal to the average value of $x(t)$ over one period. This is determined by

$$a_0 = \frac{1}{T} \int_0^T x(t)\,dt$$

The remaining coefficients, a_n and b_n, are evaluated for $n = 1, 2, 3, \ldots$ by

$$a_n = \frac{2}{T} \int_0^T x(t) \cos n\omega_0 t/dt$$

and

$$b_n = \frac{2}{T} \int_0^T x(t) \sin n\omega_0 t/dt$$

For the particular case of the square wave in Fig. 2–1a, the Fourier coefficients can be evaluated to give the following series:

Fourier series for the square wave in Fig. 2–1a, where $\omega_0 = 2\pi f_0$, is

$$x(t) = \frac{4V}{\pi} \left(\cos \omega_0 t - \frac{1}{3} \cos 3\omega_0 t + \frac{1}{5} \cos 5\omega_0 t - \frac{1}{7} \cos 7\omega_0 t + \ldots \right)$$

In this series, $4/\pi$ is a constant resulting from the integration, and V is the peak voltage of the square wave. Also, notice that this series contains only cosine terms. This is because of the square wave's symmetric arrangement about time zero (more about symmetry later).

The series above is a complete description of the frequency content in Fig. 2–1a. From it, diagrams of both the magnitude spectrum and the phase spectrum can be constructed according to the conventions discussed in Chapter 1. These diagrams are shown in Fig. 2–2.

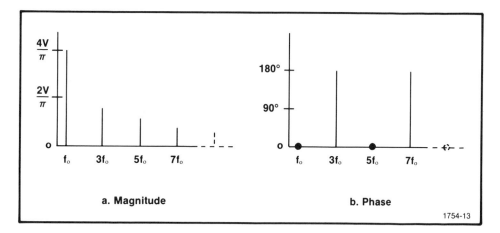

Figure 2–2   Magnitude and phase spectra for the square wave in Fig. 2–1a.

Spectral Diagrams. For most people, nature's own rainbow is the first experience with a spectral diagram. The rainbow, however, was never really thought of as a "spectrum" until Newton adapted that term in his 1672 paper to the Royal Society. This paper described the continuous bands of color produced by passing light through a prism.

More than a century after Newton's paper, Joseph von Fraunhofer (1787–1826) used diffraction gratings to study light spectra. What he saw were distinct lines instead of the continuous color spectrum observed by Newton. Von Fraunhofer also discovered that the sun and stars each have distinct spectra associated with the light they emanate. Still later, in the mid-1800s, Gustav Kirchhoff (1824–1887) and Robert W. Bunsen (1811–1899) carried von Fraunhofer's work further. (If you're in electronics, you'll recall Kirchhoff's fundamental circuit laws. And Bunsen? Well, what chemistry or physics lab is complete without a Bunsen burner?) Kirchhoff and Bunsen found that each chemical element, when heated to incandescence, radiates its own distinct color of light. Hence, each chemical element is distinguishable by a unique line spectrum. Kirchhoff used these findings to analyze the chemical composition of various unknown substances. Thus, Bunsen and Kirchhoff launched the fundamentals of general spectrum analysis.

Fourier analysis is spectrum analysis. Instead of light, though, Fourier analysis operates on waveshapes. And like light from heated compositions, different waveshapes have different spectra. For example, the square wave in Fig. 2–1a has the line spectrum shown in Fig. 2–2. That square wave is made up of sinusoids having specific frequencies with specific amplitudes and phases. No other waveshape has the same component relationships!

The spectra in Fig. 2–2 are referred to as *discrete* or *line spectra*. This is because each spectral component is discretely located on a frequency axis—each component is represented by a single line or impulse. The length of each spectral line indicates either magnitude or phase, depending on which quantity is being considered.

In the case of Fig. 2–2a, the line spectrum for magnitude is constructed from the Fourier series by first plotting the amplitude of the fundamental frequency. The fundamental frequency ($f_0 = \omega_0/2\pi$) is the reciprocal of the waveform's period and is indicated in the Fourier series by ω_0. The magnitude of the fundamental is given by the first trigonometric term in the series ($n = 1$). For the square wave example discussed thus far, the fundamental magnitude is $4V/\pi$. The diagram in Fig. 2–2a is constructed by first placing the fundamental spectral line at f_0 and giving it an amplitude of $4V/\pi$. Then subsequent Fourier terms are plotted in the same manner.

Each Fourier term is some integer multiple of the fundamental frequency and is referred to as a *harmonic*. The fundamental is sometimes referred to as the *first harmonic* because f_0 is multiplied by one, but integer multiples greater than one are always referred to as *harmonics*. In the case of Fig. 2–2a, the square wave is made up of odd harmonics. These are shown with spectral lines at $3f_0$, $5f_0$, $7f_0$, . . . , nf_0. The harmonic magnitudes are given by the Fourier coefficients in the series, and, for the square wave, they are $1/3$, $1/5$, $1/7$, . . . , $1/(2n - 1)$ of the fundamental magnitude.

The line spectrum for phase is constructed in nearly the same manner as that for magnitude. There is a spectral line in the phase diagram for each component shown in the magnitude diagram, and these are placed on a frequency axis in the same manner. The difference is that the lengths of the lines in the phase spectrum indicate phase instead of magnitude. In Fig. 2–2b, the fundamental and the fifth harmonic are positive cosines and have zero phase. Their spectral line lengths are zero, so a heavy dot is used here to indicate the presence of these zero-phase components. The third and seventh harmonics are negative cosines and therefore have phases of 180°. This is indicated by the lengths of the spectral lines at $3f_0$ and $7f_0$.

Gibb's Phenomenon. If each cosine component from Fig. 2–2 is plotted against time and the waveforms are added, the sum approximates the original square wave. This is shown in Fig. 2–3a and b.

The exact square wave is not regained in Fig. 2–3b for two reasons. First, the total number of frequency components is not added in. And second, something called *Gibb's phenomenon* (named for J. Willard Gibbs, 1839–1903, the first to study and define the phenomenon) is happening.

Let's look at the number of components used first. Only those components indicated in Figs. 2–2 and 2–3a are used to get Fig. 2–3b. The Fourier series, however, specifies that an ideal square wave contains odd harmonics out to infinity. Since it is impossible to reconstruct an ideal waveform by physically summing an infinite number of components, a lesser number is used in Fig. 2–3. This should not be too disconcerting, though. The idea of cutting off or truncating a series is really not foreign at all. Few would dispute 0.3333 as a legitimate approximation for $\frac{1}{3}$. But we all know that the decimal equivalent of $\frac{1}{3}$ is really an infinite series of threes.

If more accuracy is desired in representing $\frac{1}{3}$, 0.3333333333 can be used. For even more accuracy, even more threes can be added. This same idea is true for the Fourier series. For more accuracy, more Fourier terms can be used. This is shown in Fig. 2–3c, where the first 20 odd harmonics are summed to get more accuracy in representing the ideal square wave. But even at this, the deviation from ideal is obvious.

In short, the original waveform can never be regained exactly unless all of the terms for its Fourier series are added in. Even at that, some waveforms still cannot be regained exactly. Those that cannot be regained exactly are the types with instantaneous transitions (discontinuities).

When discontinuities exist in the original waveform, adding up its Fourier terms does provide the exact original at every point except at the discontinuities. At each discontinuity, there will always be overshoot. This overshoot is referred to as Gibb's phenomenon, and it is always equal to 8.95% of the discontinuity amplitude.

Gibb's phenomenon and the effects of truncating the Fourier series are quite

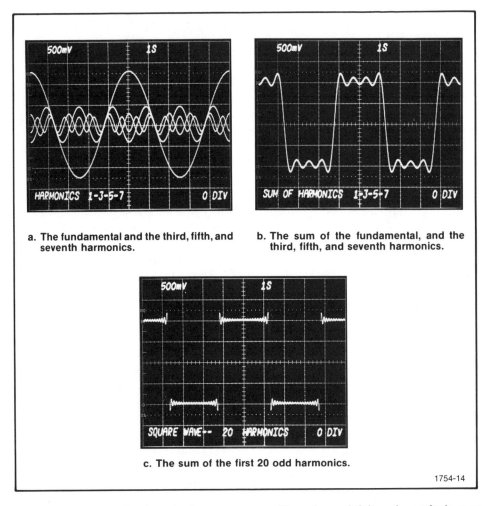

a. The fundamental and the third, fifth, and seventh harmonics.

b. The sum of the fundamental, and the third, fifth, and seventh harmonics.

c. The sum of the first 20 odd harmonics.

1754-14

Figure 2–3   Truncated Fourier series for a square wave: Truncating an infinite series results in some error in representation.

apparent in the square wave examples of Fig. 2–3. In both cases, the maximum overshoot is constant at 8.95% of the original discontinuity. The ringing that follows, however, increases in frequency and decays quicker as more series terms are added. If all the terms are added in, the picture looks like Fig. 2–4—the ringing disappears, but Gibb's phenomenon still occurs.

As in using 0.3333 to represent $\frac{1}{3}$, the key thing to realize here is that anything less than the full series is simply an approximation. If you know what to expect from truncating a Fourier series or what to expect at discontinuities, you can avoid being surprised by the results.

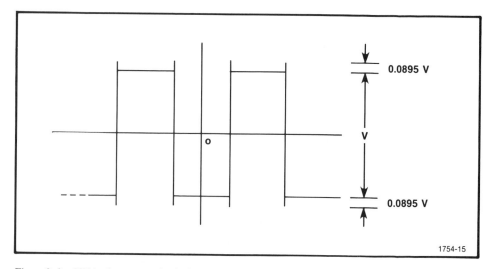

Figure 2–4   Gibb's phenomenon in the limit. Overshoot is 8.95% of the discontinuity and never disappears.

USING THE FOURIER SERIES— MATTERS OF PRACTICALITY

A square wave is the only waveform that has been discussed so far. Its Fourier series has been given, and if the period of any square wave of interest is known, this can be used with the series to determine its exact spectral components. It is simply a matter of taking the reciprocal of the period to get the fundamental, f_0. Then this is used in the series to find the frequency and magnitude of each component.

But the world is not made exclusively of square waves. What about other waveforms?

As long as you can mathematically describe a periodic waveform as a function of time, $x(t)$, that meets the Dirichlet conditions, its Fourier series can be written. However, this may not always be easy in practice. Fortunately, though, many textbooks contain Fourier series for most common waveshapes. Much work and agony can be saved by referring to them.

For your convenience, some common waveforms and their Fourier series are given in Table 2–1. Real-life waveforms rarely fit these tabulated waveshapes exactly. Square waves with zero rise time cannot be realistically generated, and some nonsymmetries and distortions from ideal invariably occur. But, if the actual waveform approximates the ideal fairly closely, the Fourier series for the ideal waveform can give some good estimates of the frequency spectrum.

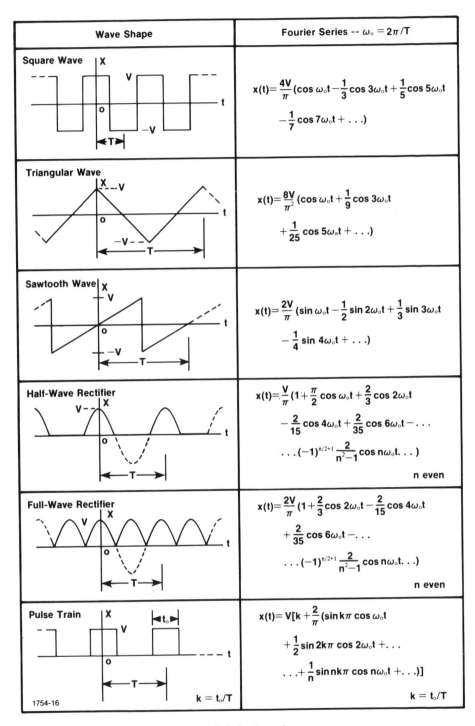

Wave Shape	Fourier Series -- $\omega_0 = 2\pi/T$
Square Wave	$x(t) = \dfrac{4V}{\pi}(\cos \omega_0 t - \dfrac{1}{3}\cos 3\omega_0 t + \dfrac{1}{5}\cos 5\omega_0 t$ $- \dfrac{1}{7}\cos 7\omega_0 t + \ldots)$
Triangular Wave	$x(t) = \dfrac{8V}{\pi^2}(\cos \omega_0 t + \dfrac{1}{9}\cos 3\omega_0 t$ $+ \dfrac{1}{25}\cos 5\omega_0 t + \ldots)$
Sawtooth Wave	$x(t) = \dfrac{2V}{\pi}(\sin \omega_0 t - \dfrac{1}{2}\sin 2\omega_0 t + \dfrac{1}{3}\sin 3\omega_0 t$ $- \dfrac{1}{4}\sin 4\omega_0 t + \ldots)$
Half-Wave Rectifier	$x(t) = \dfrac{V}{\pi}(1 + \dfrac{\pi}{2}\cos \omega_0 t + \dfrac{2}{3}\cos 2\omega_0 t$ $- \dfrac{2}{15}\cos 4\omega_0 t + \dfrac{2}{35}\cos 6\omega_0 t - \ldots$ $\ldots (-1)^{n/2+1}\dfrac{2}{n^2-1}\cos n\omega_0 t \ldots)$ n even
Full-Wave Rectifier	$x(t) = \dfrac{2V}{\pi}(1 + \dfrac{2}{3}\cos 2\omega_0 t - \dfrac{2}{15}\cos 4\omega_0 t$ $+ \dfrac{2}{35}\cos 6\omega_0 t - \ldots$ $\ldots (-1)^{n/2+1}\dfrac{2}{n^2-1}\cos n\omega_0 t \ldots)$ n even
Pulse Train	$x(t) = V[k + \dfrac{2}{\pi}(\sin k\pi \cos \omega_0 t$ $+ \dfrac{1}{2}\sin 2k\pi \cos 2\omega_0 t + \ldots$ $\ldots + \dfrac{1}{n}\sin nk\pi \cos n\omega_0 t + \ldots)]$ $k = t_0/T$

1754-16

Table 2–1   Some common waveforms and their fourier series.

Chapter 3

Fourier Integral Yields Spectra for Nonperiodic Waveforms

The Fourier series is a useful tool for investigating the spectrum of a periodic waveform, but the world is not made up exclusively of periodic waveforms. What about nonperiodic waveforms—those that don't repeat themselves in a regular fashion? Surely they too must have a frequency spectrum. After all, a bolt of lightning is nonperiodic, and certainly you've heard the familiar splatter of its spectrum on a common radio receiver.

Indeed, nonperiodic waveforms do have various frequency components. And the Fourier integral is the tool used to investigate the frequency spectra of nonperiodic waveforms.

THE FOURIER INTEGRAL IS RELATED TO THE FOURIER SERIES

The Fourier series and the Fourier integral, as analysis tools, are separate and distinct. One is intended for use with periodic waveforms and the other for use with nonperiodic waveforms. Thus, it is tempting to plunge directly into the Fourier integral since the series has already been covered. But this would be a disservice, for a subtle relationship exists between the two that is useful in interpreting later analysis concepts.

So let's not leave the Fourier series just yet. Instead, let's draw the integral out of the series by considering a periodic waveform whose period is allowed to approach infinity. Though this development of the integral may not be considered rigorous, it is certainly enlightening. And enlightenment is the goal!

Getting the Integral out of the Series. This is going to take some mathematics. However, wandering through a seemingly endless string of equations is not the answer either. Actually, there are only four major steps or equations that one needs to be aware of. These are clearly numbered below. Some information about the transitions between these steps is also given. The full details are not given. But, for later study, the complete exercise is included in many textbooks, such as, *Network Analysis* by M. E. Van Valkenburg and *Basic Network Theory* by Paul M. Chirlian (both listed in the Bibliography), to name two good examples.

To begin, let's restate the general form of the Fourier series:

[1]
$$x(t) = a_0 + \sum_{n=1}^{\infty} (a_n \cos 2\pi n f_0 t + b_n \sin 2\pi n f_0 t)$$

More appropriate to developing the integral, this should be re-expressed in exponential form. This can be done by first expressing $\cos 2\pi n f_0 t$ as

$$\left(e^{j2\pi n f_0 t} + \frac{e^{-j2\pi n f_0 t}}{2} \right)$$

and $\sin 2\pi n f_0 t$ as

$$\left(e^{j2\pi n f_0 t} - \frac{e^{-j2\pi n f_0 t}}{2j} \right)$$

where e is the base of the natural logarithm and j is the imaginary unit of the complex number system $j = \sqrt{-1}$. By some further manipulation and assignment of new variables, the more compact exponential form of the Fourier series is reached.

[2]
$$x(t) = \sum_{n=-\infty}^{\infty} c_n e^{j2\pi n f_0 t}$$

where c_n is evaluated for $n = -\infty, \ldots, -2, -1, 0, 1, 2, \ldots, \infty$ by

$$c_n = \frac{1}{T} \int_{T/2}^{T/2} x(t) e^{-j2\pi n f_0 t} dt$$

For each n, c_n is evaluated to give the magnitude and phase of the harmonic component of $x(t)$ having frequency $n f_0$.

With the Fourier series in exponential form, the next step is to recognize that each harmonic is separated by an amount $\Delta f = 1/T$. Now, with some further manipulations, which are again omitted for the sake of brevity, the series can be placed into a form that allows inspection of the limit as T goes to infinity.

[3]
$$x(t) = \lim_{T \to \infty} \frac{1}{T} \sum_{n=-\infty}^{\infty} X(n f_0) e^{j2\pi n f_0 t}$$

but since $\Delta f = 1/T$,

$$x(t) = \lim_{\Delta f \to 0} \sum_{n=-\infty}^{\infty} X(n f_0) e^{j2\pi n f_0 t} \Delta f$$

From here, as Δf goes to zero (period, T, goes to infinity), the properties of the summation approach those of an integral.

All of this simply says that when the time period, T, becomes infinity, the Fourier series reduces to

[4]
$$x(t) = \int_{-\infty}^{\infty} X(f)e^{j2\pi ft}df$$

The Fourier coefficients now have become a function of a continuous frequency variable, f, and are given by

$$X(f) = \int_{-\infty}^{\infty} x(t)e^{-j2\pi ft}dt$$

These two integrals, together, are referred to as the *Fourier transform pair*. The former, [4], is generally referred to as the *inverse Fourier transform* and the latter as the *direct Fourier transform*, or simply the *Fourier transform*.

The Fourier Integral as a Transform. As implied by its development, the Fourier integral is only applicable to nonperiodic waveforms—waveforms of infinite period. (An infinite period simply implies that the waveform does not repeat itself.)

A nonperiodic waveform, given by $x(t)$ and subject to the Dirichlet conditions, can be transformed from a function of time to a function of frequency by using

$$X(f) = \int_{-\infty}^{\infty} x(t)e^{-j2\pi ft}dt$$

When this is done, $X(f)$ is generally referred to as *the Fourier transform of* $x(t)$. And in the same manner, using

$$x(t) = \int_{-\infty}^{\infty} X(f)e^{j2\pi ft}df$$

the frequency-domain function, $X(f)$, can be *inverse transformed* back to the time domain function, $x(t)$.

Dirichlet Conditions for Transform Existence

In order for $x(t)$ to be transformed by the Fourier integral, $x(t)$ must be nonperiodic. In addition to this, the following Dirichlet conditions must be met for existence of the transform.

1. For $-\infty \leq t \leq \infty$, $x(t)$ must contain a finite number of maxima and minima.
2. If $x(t)$ contains discontinuities, they must be finite in number over the range $-\infty \leq t \leq \infty$.
3. The function $x(t)$ must be integrable in the sense that

$$\int_{-\infty}^{\infty} |x(t)|dt < \infty$$

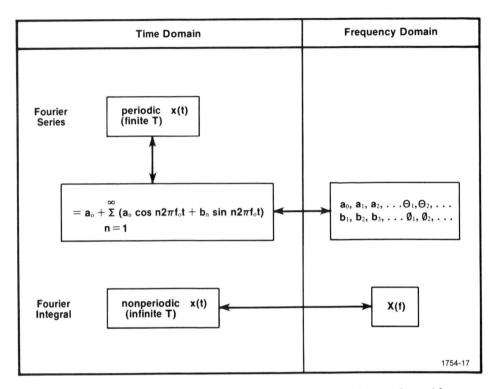

Figure 3–1   The Fourier series and the Fourier integral—two different paths between time and frequency.

For all practical purposes, these conditions are met by any nonperiodic $x(t)$ that can be physically generated.

The transform action of the Fourier integral, as compared with the Fourier series, is shown in Fig. 3–1. The major thing to notice there is that each technique applies to a different class of waveform. Also, notice that each provides a different kind of frequency-domain or spectral description. The Fourier series transforms time-domain functions to frequency-domain magnitudes and phases at specific, discrete frequencies. The Fourier integral, on the other hand, evaluates to a continuous function of frequency. To look into this a little further, let's go back to the Fourier series again and let the period go to infinity. But this time, let's do it with just pictures—no mathematics.

Infinite T Causes a Continuous Spectrum.   To see how the Fourier integral arrives at a continuous frequency spectrum, let's start with the Fourier series and a periodic waveform. In particular, consider a train of square pulses arranged so that pulse width is exactly one-half the period, T. This is shown in Fig. 3–2a. The magnitude spectrum for this pulse train is also shown in Fig. 3–2a. It was con-

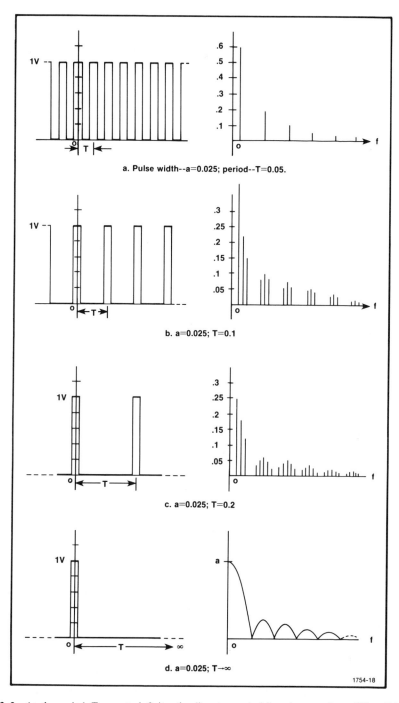

a. Pulse width--a=0.025; period--T=0.05.

b. a=0.025; T=0.1

c. a=0.025; T=0.2

d. a=0.025; T→∞

Figure 3–2 As the period, T, goes to infinity, the discrete spectral lines become closer. When T becomes infinity, the spacing between lines is zero and a continuous spectrum exists.

structed from the Fourier series of the pulse train, and, except for a DC component, the magnitude spectrum is the same as those shown for previous square wave examples.

Next, in Fig. 3–2b, the period of the pulse train is doubled while the pulse width is held constant. The effect on the magnitude spectrum is two additional frequency components around each of the original components from Fig. 3–1a, except for the DC component.

Also, notice that the amplitude of each component in Fig. 3–2b is decreased from what appeared in Fig. 3–2a. This comes from a reduced duty factor (pulse width over period), which causes a reduction of average waveform energy over the period. Since the average waveform energy is reduced, it is natural that a reduction be seen in the magnitude spectrum—the individual components need not contribute as much to the total waveform.

In Fig. 3–2c, the period is increased again and the pulse width still held constant. Note what is happening to the magnitude spectrum.

More components are packed into the magnitude spectrum, and their amplitudes are decreased. Further increases in the period cause even more spectral components to occur with even closer spacing. As the period goes to infinity (Fig. 3–2d), the spacing between components goes to zero. In other words, the series converges to the Fourier integral. The resulting magnitude spectrum is called a *continuous spectrum* because it is defined at every frequency—there is zero spacing between frequency components. This implies that a single pulse is made up of an infinite number of sinusoidal components.

A similar demonstration for the phase spectrum could be done and would ultimately reveal the same thing—phase seeming to approach a continuous function of frequency too. It should be noted, however, for the pulse train, that phase will jump between 0° and +180°. This is because the pulse train is made up of positive and negative cosine waves (see Table 2–1, Fourier series). A positive cosine wave has zero phase, and a negative cosine wave can have either +180° or −180° of phase. With increasing pulse train period (or decreasing duty factor) more cosine waves appear, with the positive and negative ones appearing in groups. With the period at infinity, phase becomes continuous and alternates, in the fashion of a square wave, between 0° and +180°.

UNDERSTANDING FREQUENCY-DOMAIN DIAGRAMS

Frequency-domain diagrams are the key to understanding Fourier analysis. They are the means for exposing the subtle nuances of the mathematics—without them, analysis is reduced to dreary comparisons of formulas and numbers.

Frequencies Can Be Negative or Positive.   People are pretty comfortable in the time domain. So the idea of negative time isn't too unsettling. It's fairly easy to picture yesterday as negative time, right now as time zero, and tomorrow as positive time. Also, our language is very generous in supplying terms for supporting this concept. Then, now, later, past, present, future—these are all familiar terms and picturable in a variety of ways.

What about negative frequency? That's not quite as comfortable. There really aren't any other words for it. For instance, −40 Hz does not conjure up any picture different than +40 Hz. Nevertheless, the mathematics of the Fourier integral require introduction of negative frequencies. Look again at the transform pair

$$X(f) = \int_{-\infty}^{\infty} x(t)e^{-j2\pi ft}dt$$

$$x(t) = \int_{-\infty}^{\infty} X(f)e^{j2\pi ft}df$$

Both $X(f)$ and $x(t)$ are defined over frequencies and times from minus infinity to plus infinity.

Actually, the concept of negative frequency isn't any more difficult than that of negative time as long as a few basic ideas are kept in mind. These ideas can be explored with the aid of Fig. 3–3.

Figure 3–3 shows a sine wave that is periodic in the time domain from minus infinity to plus infinity. Also shown in Fig. 3–3 are several spectral diagrams for this sine wave.

Consider the uppermost pair of diagrams. This pair contains the magnitude and phase information that would be obtained by writing the Fourier series for the sine wave. Notice that the amplitude, V, of the time-domain sine wave is reflected exactly in the magnitude spectrum. Also, the positioning of the time-domain sine wave relative to time zero indicates a delay of −90°. This same delay is also shown in the phase diagram. (Since this format was covered thoroughly in Chapter 1, it should be familiar to you. It should also be noted that the Fourier series can also be expressed in an exponential form, which will yield a negative frequency domain.)

Now let's look at the lower pair of diagrams in Fig. 3–3. These, also, are spectral diagrams. However, they describe the sine wave in the frequency domain of the Fourier transform. In theory, a periodic waveform extending from minus infinity to plus infinity cannot be transformed to the frequency domain by the Fourier integral. But, for the sake of illustration, this restriction is ignored momentarily. The lower pair of diagrams show what the frequency domain would look like if an infinite extent sine wave could be Fourier transformed. If this breach of theory bothers you, think of the sine wave as being the only illustrated component of some arbitrary pulse. Then its transformation, as part of the pulse, is in concert with theory.

Setting aside any remaining hesitancy about theory, go ahead and look closely at the lower pair of diagrams. Compare them with the diagrams of the series spectrum. The first difference you'll probably notice is the presence of negative frequencies in the frequency-domain diagrams. This should be expected, since the transform integral defines both positive and negative frequencies.

Now focus your attention solely on the frequency-domain magnitude diagram. Notice that there are two spectral components there, one at the positive frequency

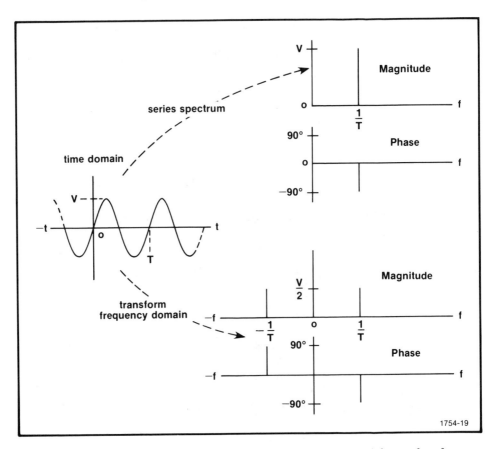

Figure 3–3 Comparison of diagramming conventions in the series spectrum and the transform frequency domain.

of the sine wave and one at its negative frequency. Also, notice that each magnitude is one-half that of the time-domain waveform (also, each is half the magnitude of the one in the series spectrum). Since the spectrum is divided between positive and negative frequencies, doesn't it seem reasonable to divide the energy in the same manner?

Now let's look at the frequency-domain phase diagram. The phase for positive frequency duplicates the phase shown in the series spectrum; it is −90° at $1/T$. Phase in the negative frequency domain is simply an inverted image of the positive domain. For this example, it is +90° at $−1/T$.

Are you wondering why phase isn't halved like magnitude? It's because phase is just a position indicator, not an energy indicator.

The frequency-domain diagrams in Fig. 3–3 embody most of the conventions of magnitude and phase description for the Fourier transform's frequency domain. In review, these conventions are

1. *The magnitudes in the positive and negative frequency domain exactly duplicate each other.* Except for the DC component, the magnitudes are equally divided between the positive and negative frequency domain. For every frequency indicated in the positive frequency domain, one of equal magnitude is indicated at the same frequency in the negative domain. Their sum equals the amplitude of the corresponding sinusoidal component in the time domain. In the case of DC, its frequency is zero and there is no division of magnitude.

   There is one qualification that should be noted here. The convention above is true only for real-valued signals. In the case of complex signals, those of the form $x(t) = a(t) + jb(t)$, the negative frequency domain will not mirror the positive frequency domain. However, the majority of signals in actual measurement situations are real. For them, the negative frequency domain does mirror the positive frequency domain.

2. *Phase in the positive frequency domain is duplicated in negative frequency, except the images are inverted.* This simply entails a sign change when passing between positive and negative frequencies. The amount of phase in the time domain is reflected exactly the same in both the positive and negative frequency domain. Unlike magnitude, phase is not halved.

   Even with these conventions in mind, you may still be wondering; "Just exactly what does a negative frequency sine wave look like?"

   Well, let's take a look at one. Figure 3–4 shows a three-dimensional space that can be associated with the Fourier transform. This is an extension of the three-dimensional pictorial aid presented in Figs. 1–4 and 1–8 in Chapter 1. The difference in Fig. 3–4 is that negative frequencies are shown and frequency-domain conventions are used.

   In keeping with frequency-domain conventions, two sinusoids are shown in Fig. 3–4. One is the negative frequency term of a time-domain component and the other is the positive frequency term. Each sinusoid is of equal amplitude, and each passes through the frequency axis at points equal to $-1/T$ and $+1/T$. Their positive amplitude projections onto the frequency amplitude plane at these two points form the frequency-domain magnitude diagram. Also, their full projection onto the amplitude time plane is summed to equal the time-domain component they represent (shown by a dashed line in Fig. 3–4). In order for this to occur, both the positive and negative frequency terms must have the same phase, and in Fig. 3–4 they do. The positive frequency term is arranged for $-90°$ phase (a sine wave). The corresponding term in negative frequency has exactly the same arrangement with time; however, it is said to have $+90°$ phase. This sign change is in accordance with the conventions of frequency-domain description and is shown in the phase diagram at the bottom of Fig. 3–4.

   Thus, a negative frequency sine wave looks exactly like a positive frequency sine wave. Each is just located at points of opposite sign on the frequency axis.

   Now that some of the conventions for frequency-domain diagrams have been

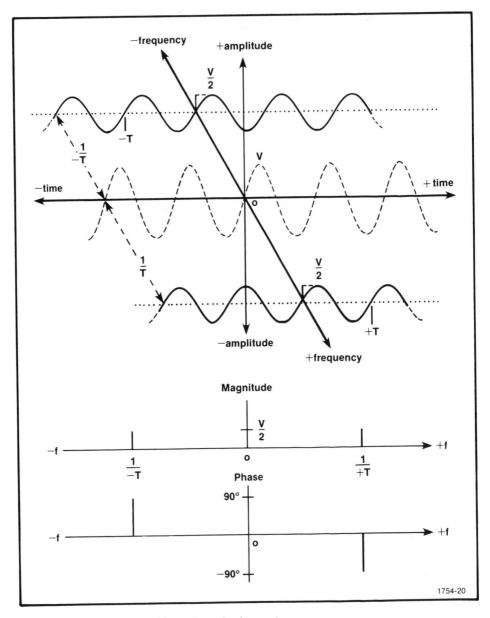

Figure 3–4 A 3-D look at positive and negative frequencies.

covered, let's go back to Fig. 3–2d and redraw it for the Fourier integral's frequency domain. This redrawn version is shown in Fig. 3–5. Notice that the time-domain pulse (Fig. 3–5a) is a more general version with an amplitude of V and a width of $2T_0$. Also, the pulse is centered about time zero, giving it symmetry about the amplitude axis.

The frequency-domain diagrams for the pulse are shown in Fig. 3–5b and c. In Fig. 3–5b, the magnitude is shown as a continuous function of frequency, and

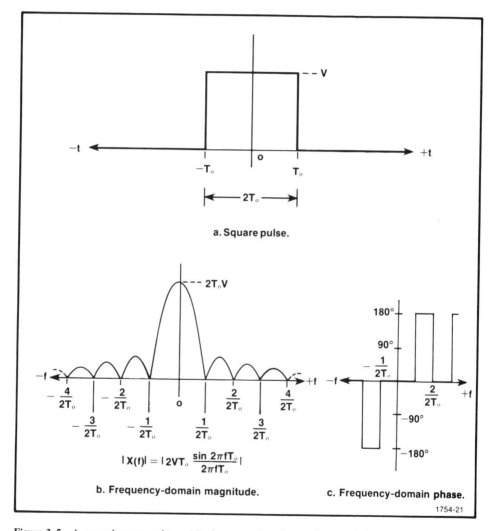

a. Square pulse.

$$|X(f)| = |2VT_0 \frac{\sin 2\pi fT_0}{2\pi fT_0}|$$

b. Frequency-domain magnitude.

c. Frequency-domain phase.

1754-21

Figure 3–5 A general square pulse and its frequency-domain magnitude and phase.

the positive-frequency magnitude is exactly mirrored in the negative frequencies. The phase, as shown in Fig. 3–5c, is continuous and alternates as a square wave between 0° and 180°. This phase arrangement indicates that the square pulse is made up entirely of cosine waves, the positive cosine waves giving rise to 0° phase and the negative ones giving rise to 180° phase. Because the spectra are continuous functions, every frequency of cosine wave is present in the pulse, except at the magnitude nulls or zeros at $\pm 1/2T_0$, $\pm 2/2T_0$, $\pm 3/2T_0$, ... For phase, these same nulls are points of transition between 0° and 180°.

As a final note, Fig. 3–5 applies to any square pulse centered around time zero. Just substitute the actual amplitude of the pulse for V and its width for $2T_0$. Both the magnitude and phase diagrams will then apply. For square pulses, not centered around time zero, the same magnitude diagram will still apply; however, phase will vary according to the pulse's shift relative to time zero.

Frequency-Domain Descriptions Can Be Rectangular in Form.   Up to this point, the frequency domain has been described only in terms of magnitude and phase. This is probably the best introductory approach since sinusoids and phase angles are generally familiar terms. Also, introduction through magnitude and phase lends itself well to various graphical explanations and descriptions.

But there is more than one way to look at the frequency domain. In fact, the actual computation involved in Fourier transformation doesn't necessarily lead directly to magnitude and phase results. It's often more direct to compute the frequency-domain results in rectangular form such that

$$X(f) = \int_{-\infty}^{\infty} x(t)e^{-j2\pi ft}dt = Re(f) + jIm(f)$$

In this relationship, the time-domain function being transformed to the frequency domain is $x(t)$, and the frequency-domain result is given as $Re(f) + jIm(f)$.

$Re(f)$ is referred to as the *real part* of the frequency domain, and $Im(f)$ is referred to as the *imaginary part*. Together, these two parts are referred to as a *complex-valued function in rectangular form*—complex because the function is of more than one part, and rectangular because complex quantities are often portrayed as vectors in rectangular coordinates. This is really no different from magnitude and phase, which comprise a function of two parts—magnitude and phase. Also, magnitude and phase can be portrayed as a rotating vector in polar coordinates. Thus, magnitude and phase are often referred to as *results in polar form*.

The relationship between the rectangular and polar forms is shown by vectors in Fig. 3–6. If you are not at all excited by vectors (I'm not), you might prefer to look at Fig. 3–7. The same relationship is shown there through the time-domain and frequency-domain functions for a single pulse.

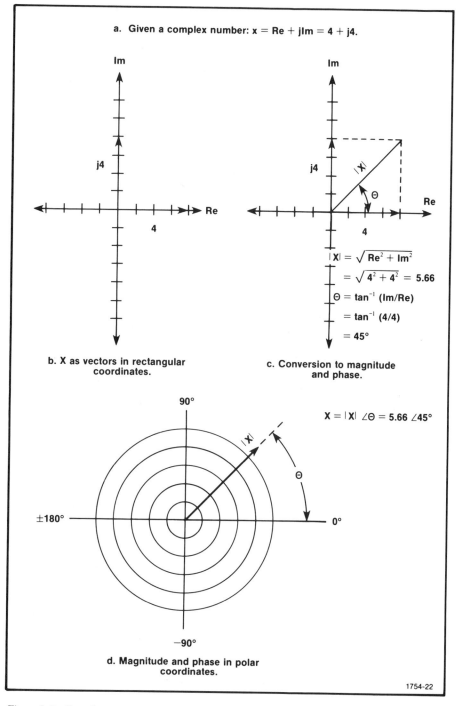

a. Given a complex number: x = Re + jIm = 4 + j4.

b. X as vectors in rectangular coordinates.

c. Conversion to magnitude and phase.

$$|X| = \sqrt{Re^2 + Im^2}$$
$$= \sqrt{4^2 + 4^2} = 5.66$$
$$\Theta = \tan^{-1}(Im/Re)$$
$$= \tan^{-1}(4/4)$$
$$= 45°$$

$$X = |X| \angle \Theta = 5.66 \angle 45°$$

d. Magnitude and phase in polar coordinates.

1754-22

Figure 3–6 Complex number expressed in rectangular and polar form.

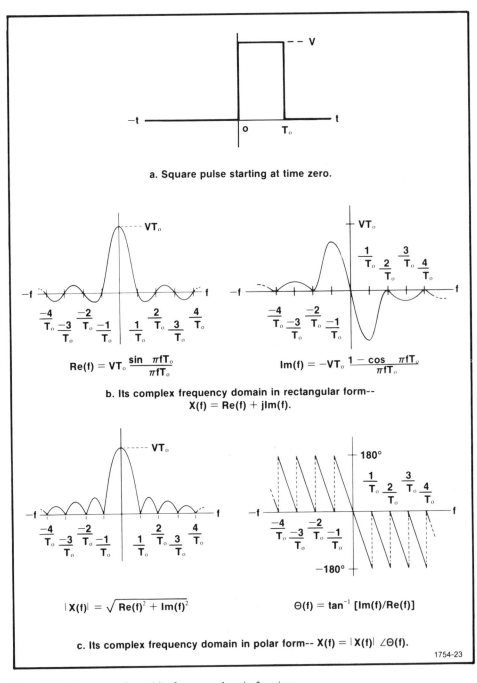

a. Square pulse starting at time zero.

$$Re(f) = VT_o \frac{\sin \pi f T_o}{\pi f T_o}$$

$$Im(f) = -VT_o \frac{1 - \cos \pi f T_o}{\pi f T_o}$$

b. Its complex frequency domain in rectangular form--
$$X(f) = Re(f) + jIm(f).$$

$$|X(f)| = \sqrt{Re(f)^2 + Im(f)^2}$$

$$\Theta(f) = \tan^{-1} [Im(f)/Re(f)]$$

c. Its complex frequency domain in polar form-- $X(f) = |X(f)| \angle\Theta(f).$

1754-23

Figure 3–7 A square pulse and its frequency domain functions.

FUNCTIONS ARE EVEN OR ODD
OR THE SUM OF EVEN AND ODD PARTS

If you can identify even and odd time-domain functions, you can predict some frequency-domain features of waveforms just by examining the waveform itself.

Even Functions Are the Sum of Cosines Only. Figure 3–7 is useful for pointing out the relationship between the rectangular and polar form of the Fourier transform. The real interest, however, arises when this square pulse is compared with the one in Fig. 3–5. These two pulses really aren't too different in the time domain. But look at their frequency domains. There are some distinct differences, especially in the shapes of the phase functions.

What's so different in the time domain that makes such a difference in the frequency domain? Well, first of all, you might notice that the pulse in Fig. 3–5 has twice the width of that in Fig. 3–7. Does this make the difference? Not really. Both of the square pulses are given in general enough form so that pulse width doesn't affect the shape of the frequency-domain function. The only effect is in scaling. Where the pulse width in Fig. 3–5 is $2T_0$, it is T_0 in Fig. 3–7. T_0 simply replaces $2T_0$ in the scaling.

If it isn't pulse width, then what makes the difference in phase? Compare the two pulses again. In particular, look at each pulse's position relative to time zero. In Fig. 3–5, the square pulse starts at some negative time and ends at an equivalent positive time. The left half of the pulse is the same as the right half. If you think about it, you can picture the two halves being folded about time zero to come together for an exact match. The pulse is symmetric about time zero. Mathematically, this is stated by the following equality:

$$x(t) = x(-t)$$

Any function, periodic or nonperiodic, that meets the $x(t) = x(-t)$ condition is said to be an *even function of time*. The same can be said for the frequency domain too. A function that meets the condition of $X(f) = X(-f)$ is an *even function of frequency*.

Table 2–1, back in the discussion of the Fourier series, contains several examples of even functions of time. All the waveforms there, except the sawtooth wave, are even functions of time. They meet the $x(t) = x(-t)$ requirement. Also, if you look at the Fourier series for each of these even functions, you should notice something else they have in common. In particular, they are all composed of cosine terms having either a positive or a negative multiplying constant. The same is true for the square pulse in Fig. 3–5. It is an even function and is composed entirely of cosine waves having either a positive or negative amplitude term. This means that phase can be either 0° or 180°, depending on the sign of each cosine term.

All even functions of time are made up entirely of cosine waves and have phase of either 0° or 180°. Or, if you prefer to think of the rectangular form, the Fourier transform of an even function of time gives a real and even function of frequency and a zero imaginary part. Phase then depends on the sign of the real part and is 0° for the positive amplitude portions of the real part and 180° for the negative portions.

Odd Functions Have Sines Only.   Now, consider the sawtooth wave in Table 2–1. It isn't an even function, but it does appear to have some kind of symmetry. In fact, it meets the condition for being an *odd function of time*. This condition is given by the following relationship:

$$x(t) = -x(-t)$$

In terms of looking at the waveform, it is odd if the positive time portion can be sign reversed and folded about time zero for an exact match. This idea of folding and matching to determine evenness or oddness is shown in Fig. 3–8.

Returning to the sawtooth wave in Table 2–1, notice that its Fourier series is made up of sine terms only. This is characteristic of all odd functions of time. They are made up entirely of sine waves and have 90° phase, either plus or minus depending on the sign of the amplitude multiplier. Or, in terms of the rectangular form, odd functions of time transform to odd and imaginary functions of frequency. The real part of the frequency domain for any odd function is always zero.

So far we haven't looked at a pulse that is an odd function of time. So, let's look at one. Figure 3–9a shows such a pulse, and you can prove its oddness for yourself by mentally going through the process shown in Fig. 3–8. The rectangular and polar forms of this pulse's frequency-domain functions are shown in Fig. 3–9b and c. Notice that the real part of the rectangular form is zero, as it should be. The imaginary part contains all of the frequency amplitude information.

Look closely at the imaginary part in Fig. 3–9b. Is it even or odd? Just like the time-domain waveform it came from, the imaginary part is odd. This is another feature of oddness and evenness. Odd functions in one domain transform to odd, imaginary functions in the opposite domain, and even functions transform to even, real functions.

Some Functions Are Part Even and Part Odd.   Let's go back to the square pulse in Fig. 3–7. If you test this pulse for evenness and oddness, you'll find that it fits neither case. But look at its frequency domain in rectangular form. Then look at the pulses in Fig. 3–5 and Fig. 3–9. Do you see how their frequency domains, except for a multiplying constant, might somehow be combined to equal the frequency domain of the pulse in Fig. 3–7?

In fact, the square pulse in Fig. 3–7 is neither even nor odd, but it is the sum of an even and an odd function. This summation is shown more clearly in Fig. 3–10 (p. 45). Notice that this frequency-domain sum is done in rectangular form. This is because sums cannot be done directly in the polar form. So, one of the advantages of rectangular form is its directness of addition and subtraction. After looking at Fig. 3–10, do you see how the even and odd parts sum to a function that is neither even nor odd?

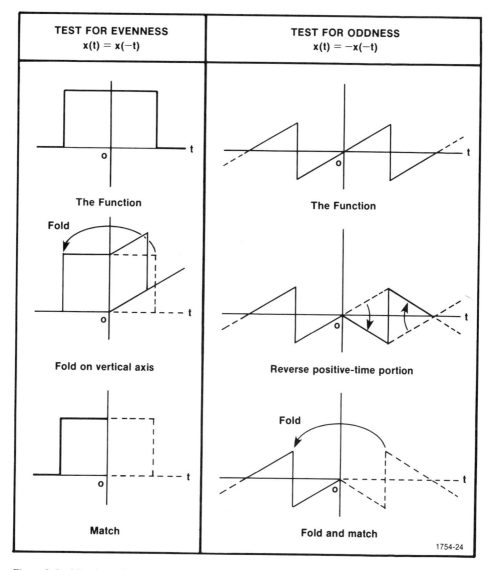

Figure 3–8 Visual test for even or odd function.

There is an important conclusion to be drawn from all of this. It is that an arbitrary function can be classed as even, odd, or the sum of even and odd parts. As you enter into Fourier analysis, keep this conclusion in mind. It can help you predict what form the results should take and thus keep you from continuing on erroneous paths.

Actually, the square pulse resulting from the summing in Fig. 3–10 can be turned back into an even function. Just remove some of its time delay by shifting it

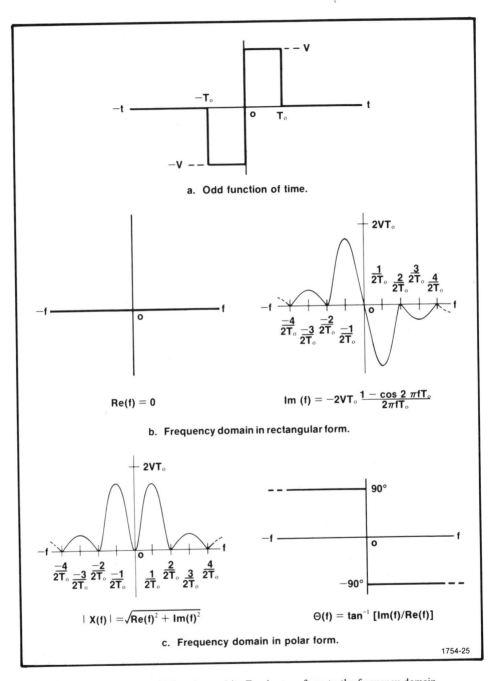

a. **Odd function of time.**

$Re(f) = 0$ $Im\ (f) = -2VT_0 \dfrac{1 - \cos 2\ \pi fT_0}{2\pi fT_0}$

b. **Frequency domain in rectangular form.**

$|\ X(f)\ | = \sqrt{Re(f)^2 + Im(f)^2}$ $\Theta(f) = \tan^{-1}\ [Im(f)/Re(f)]$

c. **Frequency domain in polar form.**

1754-25

Figure 3–9   Pulse that is an odd function and its Fourier transform to the frequency domain.

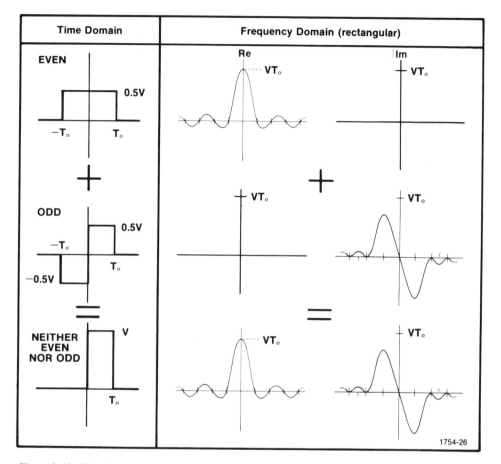

Figure 3–10 Waveforms are either even or odd or the sum of even and odd parts.

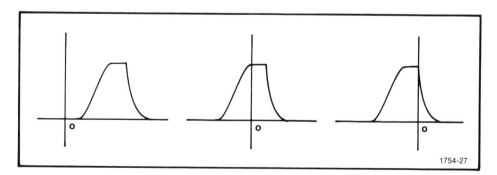

Figure 3–11 Some functions cannot be shifted for evenness or oddness.

left until it is centered on the time origin. Then, the square pulse is an even function again. Its imaginary part becomes zero, and its real part remains nonzero. In terms of the polar form, time shifting a waveform has no effect on frequency-domain magnitude and is only reflected as a change in phase.

Shifting a function to take advantage of symmetry is a standard mathematical operation. If a function can be arranged to be even or odd, both Fourier analysis and the results obtained are simplified. There are still functions, however, that cannot be shifted for evenness or oddness. One such function is shown in Fig. 3–11 (p. 45). There are many others like this in real life analyses.

PERIODIC OR NONPERIODIC?— IT'S YOUR POINT OF VIEW

Up to this point, the discussions of the Fourier series and the Fourier integral have taken two narrow points of view. Every waveform example has been defined to be either periodic (repeating itself from minus infinity to plus infinity) or nonperiodic (a transient or pulse occurring only once over infinite time). Everything has been done in accord with the strict theoretical definitions of periodic and nonperiodic. But to stick to these definitions in practice invites conflict.

Consider, for example, a "sine-wave" oscillator—an actual physical circuit. When we turn on the oscillator and look at its output with an oscilloscope, we see something that certainly looks like a sine wave and keeps repeating itself in a periodic fashion. And when the oscillator is turned off, the output ceases. Everything is working as it should. We have a circuit that generates a periodic waveform, a sine wave. Right?

Wrong! Not if we are going to stay with the purely mathematical definition of periodicity. We turned the oscillator on, watched the output repeat itself for a while, then turned it off. The oscillator's output didn't repeat itself over all time from minus infinity to plus infinity. In fact, the oscillator wasn't even built until quite some time after minus infinity. It generated a sinusoid for only a finite window in the finite continuum of time. It's as if a theoretical sine wave had been multiplied by a square pulse in the manner of Fig. 3–12.

"But," you might say, "theoretical definitions aside, it's periodic as far as I'm concerned—at least for the time I looked at it."

And that's a good point of view. It's a practical point of view. For any physically generated waveform that repeats itself—square waves, sawtooths, and so forth—you'll probably want to take the periodic point of view and use the Fourier series for analysis. All you need for the Fourier series approach is one complete cycle of the waveform. But remember, using the Fourier series does imply theoretical periodicity. You'll be treating the waveform as if it did extend forever beyond the edges of the finite time window you are actually dealing with. The analysis results will contain the discrete harmonics that make up the periodic waveform.

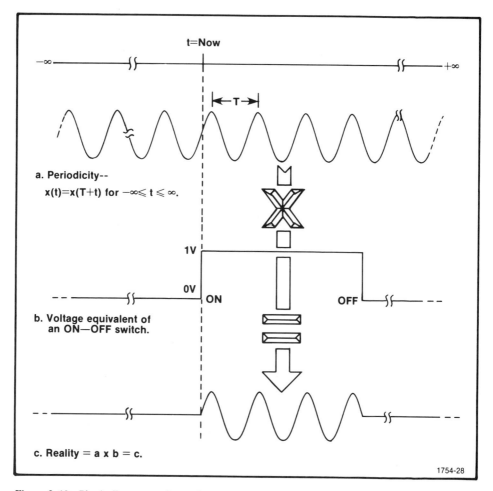

Figure 3–12 Physically generated waveforms are either pulses (b) or windowed representations (c) of theoretical waveforms.

On the other hand, you might want to take a different point of view. For example, let's say you have a sine-wave oscillator that you want to use in a remote-control application. Maybe you want to send a very short burst of the sine wave to tone control an unpiloted aircraft or a robot or simply to activate a switch remotely. Whatever the case, you'll be interested in the effects of gating the sine wave on and off (windowing). For that, you'll want to take a nonperiodic point of view and do your analysis with the Fourier integral.

To explore the nonperiodic point of view further, let's go ahead and see how the Fourier integral is applied to a windowed waveform. And maybe more importantly, let's find out what happens when a periodic waveform is windowed.

Fourier Transform of a Rectangular Window. The rectangular window that is going to be used is shown in Fig. 3–13. You'll probably recognize it immediately as our old friend, the square pulse.

You already know what the frequency domain of this rectangular window looks like (Fig. 3–5), and that will be of more interest shortly. But first, let's look at how the Fourier integral is used to obtain the window's frequency domain.

As you may recall, the integral for transformation to the frequency domain is given by

$$X(f) = \int_{-\infty}^{\infty} x(t)e^{-j2\pi ft}dt$$

For the case of the rectangular window, however, this can be somewhat simplified. To do this, notice that the window has a value of zero everywhere except over the interval from $-T_0$ to T_0. In this nonzero interval, the window has a constant amplitude of one. Since the Fourier transform of zero is zero, it doesn't make sense to apply the Fourier integral over anything but the nonzero interval. So, the integral can be changed to

$$X(f) = \int_{-T_0}^{T_0} x(t)e^{-j2\pi ft}dt$$

And since the rectangular window, $x(t)$, has a constant value of one during the interval, the integral is further reduced to

$$X(f) = \int_{-T_0}^{T_0} e^{-j2\pi ft}dt$$

This is a fairly innocuous expression for those familiar with calculus. It can be evaluated with standard textbook methods. But let's not worry about mechanics of evaluation. Let's just look at the answer.

The integral evaluates to

$$X(f) = 2T_0 \frac{\sin 2\pi fT_0}{2\pi fT_0} + j0$$

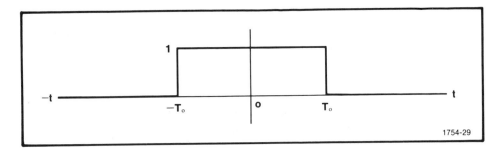

1754-29

Figure 3–13   A rectangular time window.

which is the Fourier transform in rectangular coordinates (note that the imaginary part is zero). It has exactly the same magnitude and phase, when converted to polar form, as shown earlier in Fig. 3–4. So, a rectangular window is the same in every respect as a square pulse.

Also, for later reference, notice that the form of the real part is the sine of an argument divided by the argument. This is a classic frequency-domain waveshape description, having a major lobe with decaying side lobes, and is often generalized in discussions as $(\sin x)/x$.

Rectangularly Windowed Waveforms. If you recall the initial discussion of the Fourier transform, you may remember that it doesn't exist for periodic waveforms. But that needn't stop us. We now have a window!

To transform sine waves, square waves, and other periodic waveforms, all you need to do is limit your view of the waveform. Look at it through a window. In short, just transform a selected interval of the waveform instead of the whole thing. This is done in the same manner as discussed for the square pulse or rectangular window and is illustrated in Fig. 3–14. But keep in mind that the spectra associated with a strictly periodic waveform and a windowed version of the same signal will be different. One will have a discrete spectrum (the periodic signal), and the other will have a continuous spectrum (the nonperiodic or windowed signal).

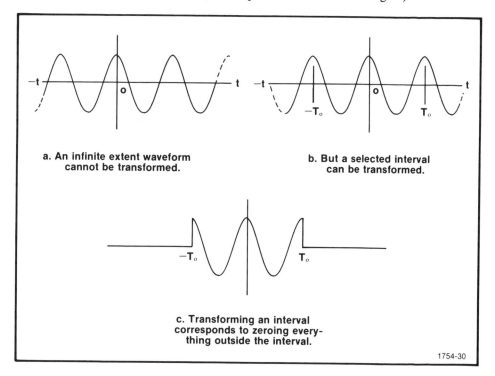

a. An infinite extent waveform
cannot be transformed.

b. But a selected interval
can be transformed.

c. Transforming an interval
corresponds to zeroing everything outside the interval.

1754-30

Figure 3–14   Transforming a selected interval can change the waveform.

In Fig. 3–14a, an infinite extent waveform is shown, in this case, a cosine wave. Since it cannot be transformed in its entirety, an interval for transformation has been marked off in Fig. 3–14b. Notice that this interval, from $-T_0$ to T_0, is the same as indicated for the rectangular window in Fig. 3–13. The act of transforming over this interval with the Fourier integral, however, implies a zero-valued waveform outside the interval. So what is really transformed is shown in Fig. 3–14c. It is as if the cosine wave was actually multiplied by the rectangular window shown in Fig. 3–13. In essence, this is what happens whenever Fourier transformation is done over a finite interval.

But what does this mean in terms of the frequency domain? Obviously, Fig. 3–14c is not the same as Fig. 3–14a. The actual process of windowing a waveform is shown in Fig. 3–15. The same cosine wave and rectangular window are used. Additionally, each step is shown in the frequency domain, so its effects there can be seen. To keep matters simple at this stage, phase is ignored and only the frequency-domain magnitude is shown.

In Fig. 3–15a, an infinite extent cosine wave is shown in the time domain and as two spectral components at frequencies $\pm 1/T$. Following it in Fig. 3–15b is a rectangular window in the time domain and its $(\sin x)/x$ magnitude in the frequency domain. Multiplying the time-domain window and the cosine wave results in the product shown in Fig. 3–15c. This product is referred to as a rectangularly windowed cosine wave, and the magnitude of its Fourier transform is shown in Fig. 3–15d.

Notice that the transform magnitude of the windowed cosine wave is not the product of the frequency-domain functions for the cosine wave and the window. Multiplication in the time domain does not correspond to multiplication in the frequency domain.

Yet, you may notice that Fig. 3–15d does exhibit some of the traits of both the cosine wave and the rectangular window. The $(\sin x)/x$ magnitude of the window appears twice in Fig. 3–15d, once in positive frequency and once in negative frequency. And the peaks of the double $(\sin x)/x$ magnitude occur at the frequency $(\pm 1/T)$ of the cosine wave. It's as if the two functions were somehow "rolled together" to give a new function having the major features of its constituents. In fact, the two functions are "rolled together" by a mathmatical process called *convolution*. Figure 3–15d is the result of convolving the cosine wave's frequency domain with the window's frequency domain.

So, multiplication in the time domain is equal to convolution in the frequency domain. And the transform of a windowed function is equal to the convolution of the function's frequency domain and the window's frequency domain.

Convolution is an interesting subject. It's one that tempts digression into pages and pages of discussion. But the purpose here is to cover the broader aspects of Fourier analysis first. Still, an understanding of convolution, if only a superficial one, is helpful background. (For a complete development of convolution, refer to M. E. Van Valkenburg's, *Network Analysis,* pp. 222–33, listed in the Bibliography.) So let's look at the operation just briefly.

Convolving two functions, given by $h(t)$ and $x(t)$, results in a third function,

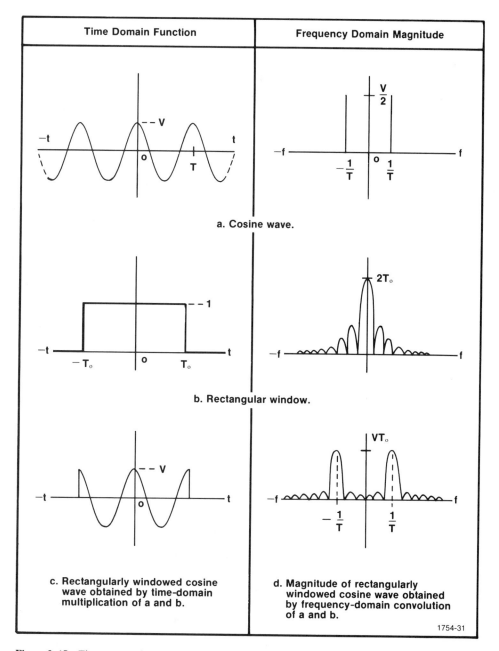

Time Domain Function	Frequency Domain Magnitude

a. Cosine wave.

b. Rectangular window.

c. Rectangularly windowed cosine wave obtained by time-domain multiplication of a and b.

d. Magnitude of rectangularly windowed cosine wave obtained by frequency-domain convolution of a and b.

1754-31

Figure 3–15   The process of rectangular windowing.

$y(t)$. The operation is further defined by the following integral, which uses τ as a dummy variable to facilitate shifting of one function past the other.

$$y(t) = \int_{-\infty}^{\infty} h(\tau)x(t - \tau)d\tau$$

Rather than dwell on this integral, Fig. 3–16 provides a graphical demonstration of convolution. As can be seen there, convolution consists of flipping one waveform over in time. Then the waveform is shifted past the other, all the while forming the product of the two and integrating for the area of the product. The result, $y(t)$, is the area of the product versus shift, t.

The same shifting-multiplying-integrating operation of convolution can be carried out in the frequency domain as well. In fact, visualizing it is necessary to understanding the Fourier transform's frequency domain. As an example, visualize the operation for an impulse (a single-valued function, or a spike with zero width) and the $(\sin x)/x$ magnitude function. Referrring back to Fig. 3–15 will help in the visualization.

First, picture the $(\sin x)/x$ magnitude in Fig. 3–15b being flipped in frequency. The result should be no change because of symmetry about zero frequency. Now picture shifting it past a single impulse (for example, the one at $1/T$ in Fig. 3–15a) and integrating the product of the two. The result is simply a tracing out or repeating of the $(\sin x)/x$ magnitude with its center of symmetry about $1/T$. If you do this

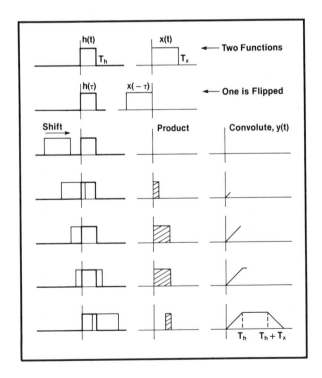

Figure   3–16 Convolution   performed graphically.

for each impulse by itself and add the results, you get what is seen in Fig. 3–15d, which is the convolution of the window's and the cosine wave's frequency domains.

The topics of windowing and convolution are touched on again in Part II of this book. For now, however, the important points to keep in mind are

1. Infinite extent waveforms (sinusoids, square waves, and so forth) can be transformed to the frequency domain by the Fourier integral.

2. To do this, however, you must multiply the waveform by a finite time window.

3. This multiplication in the time domain is equivalent to convolution in the frequency domain. So the transform of a windowed signal is the convolution of the window's and signal's individual frequency-domain representations.

As a final note, the converse of item 3 above is also true. Multiplication in the frequency domain is equivalent to convolution in the time domain.

A SUMMARY OF SOME IMPORTANT FOURIER TRANSFORM PROPERTIES

Thus far, several important properties of the Fourier transform have been covered. For example, the idea that an arbitrary waveform is made up of odd and even parts has been discussed to some extent. However, some properties have only been presented subtly in discussing other aspects of the transform, and some important properties haven't been covered at all. To remedy this, the more important properties of the Fourier transform are as follows:

1. *The Fourier transform has an inverse.* Although most of the discussion and examples have focused on transforming from the time domain to the frequency domain, a transform may also be done in the opposite direction. That is, a frequency-domain function can be transformed to obtain its corresponding time-domain function. This is the other half of the Fourier integral pair and is illustrated in Fig. 3–17.

2. *Even functions transform to real parts only.* If a real function given by $x(t)$ satisfies the relation $x(t) = x(-t)$, then $x(t)$ transforms to the real part of the frequency domain only and will be an even function of frequency. The imaginary part of its frequency domain will be zero. This property has already been discussed to some length and is illustrated in Figs. 3–8 and 3–10.

It is also worth noting that, in the polar form, phase is zero over the frequencies corresponding to the positive real part and 180° over the frequencies where the real part is negative.

3. *Odd functions transform to imaginary parts only.* If a real function given by $x(t)$ satisfies the relation $x(t) = -x(-t)$, then it transforms to the imaginary part of the frequency domain only and is an odd function of frequency. The real part is zero. This property has already been discussed and is illustrated by Figs. 3–8, 3–9, and 3–10.

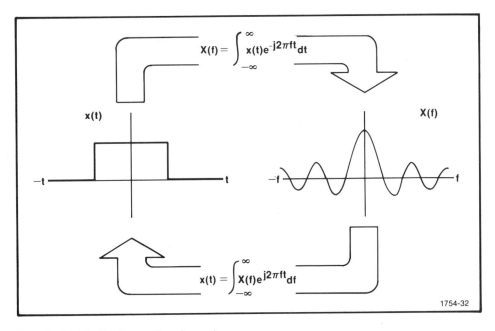

Figure 3–17 The Fourier transform has an inverse.

In the case of phase, it will be either $+90°$ or $-90°$, depending on the sign of the imaginary part at any frequency.

4. *Arbitrary functions are the sum of even and odd parts.* Any function may be expressed as the sum of even and odd parts. In an even function, the odd part is zero. In an odd function, the even part is zero. Where a function is neither even nor odd, it is the sum of nonzero even and odd parts, and its frequency domain has a real and an imaginary part, both of which are nonzero. This concept is summed up by Fig. 3–10.

5. *A component added in time is a component added in frequency (linearity property).* Suppose you have two functions given by $x(t)$ and $y(t)$ and they Fourier transform to $X(f)$ and $Y(f)$. Then, $x(t) + y(t)$ transforms to $X(f) + Y(f)$. This is most often referred to as the *linearity property*, or the *property of being a linear transform*.

The groundwork for this property was laid in Chapter 1 and is illustrated there by Figs. 1–7 and 1–8. Also, it is implied in the discussion of summing even and odd parts and is further illustrated in Fig. 3–10.

6. *Time scaling affects frequency and amplitude scaling.* Suppose you have a function given by $x(t)$, and it Fourier transforms to $X(f)$. Now suppose you wish to rescale $x(t)$ in time by a factor k, where k is a nonzero constant. Then $x(kt)$ Fourier transforms to $X(f/k)/|k|$.

In other words, a time-scale expansion corresponds to a frequency-scale compres-

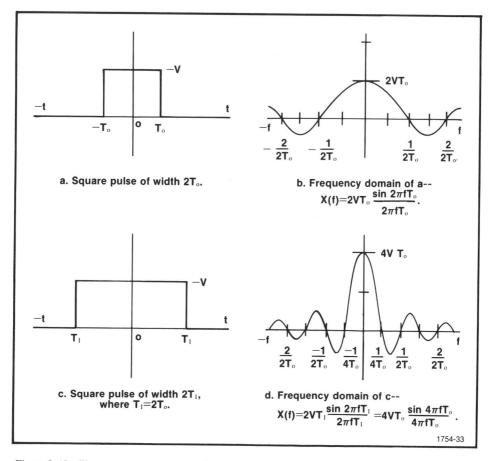

a. Square pulse of width 2T₀.

b. Frequency domain of a--
$$X(f)=2VT_0\frac{\sin 2\pi fT_0}{2\pi fT_0}.$$

c. Square pulse of width 2T₁, where T₁=2T₀.

d. Frequency domain of c--
$$X(f)=2VT_1\frac{\sin 2\pi fT_1}{2\pi fT_1}=4VT_0\frac{\sin 4\pi fT_0}{4\pi fT_0}.$$

1754-33

Figure 3–18 Time expansion compresses frequency, increases amplitude.

sion and increased frequency-domain amplitude. A time-scale compression causes the opposite to occur—an expanded frequency scale and decreased amplitude. This time-scaling property is more clearly demonstrated in Fig. 3–18. There, the scaling is constant, and the property is demonstrated through variation of the pulse-width parameter.

7. *Frequency scaling affects time and amplitude scaling.*   This property is comparable to time scaling. Suppose you have a function given by $x(t)$ and it is Fourier transformed to $X(f)$. Now suppose you wish to rescale $X(f)$ in frequency by factor k, where k is a nonzero constant. Then $X(kf)$ inverse transforms to $x(t/k)|k|$. This is more clearly demonstrated by Fig. 3–19.

8. *Time shifting affects phase only.*   Suppose you have a function given by $x(t)$ that Fourier transforms to $X(f)$. Now, what happens if you shift $x(t)$ by a constant time, T?

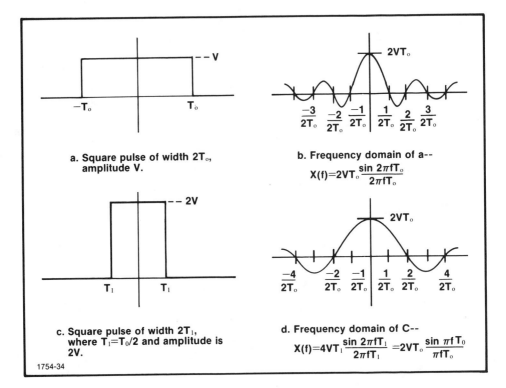

a. Square pulse of width $2T_0$, amplitude V.

b. Frequency domain of a--

$$X(f)=2VT_0\frac{\sin 2\pi fT_0}{2\pi fT_0}$$

c. Square pulse of width $2T_1$, where $T_1=T_0/2$ and amplitude is 2V.

d. Frequency domain of C--

$$X(f)=4VT_1\frac{\sin 2\pi fT_1}{2\pi fT_1}=2VT_0\frac{\sin \pi fT_0}{\pi fT_0}$$

1754-34

Figure 3–19   Frequency expansion compresses time, increases amplitude.

When you shift $x(t)$ by a constant time, T, $x(t)$ becomes $x(t - T)$, which Fourier transforms to

$$X(f)e^{-j2\pi fT}$$

This is demonstrated in Fig. 3–20. Notice there that time shifting affects phase only; magnitude remains constant throughout.

9. *Frequency shifting causes time-domain modulation.*   If $X(f)$ is inverse transformable to an $x(t)$, then shifting $X(f)$ by some constant frequency, F, results in an $X(f - F)$ that inverse transforms to

$$x(t)e^{j2\pi tF}$$

This corresponds to a cosinusoid being modulated by $x(t)$ in the time domain.

Figure 3–21 demonstrates this shifting property. Only the real part of the frequency domain is dealt with in Fig. 3–21, so the resulting time-domain functions are even. Notice that the frequency of the cosine wave equals that of the frequency shift, F.

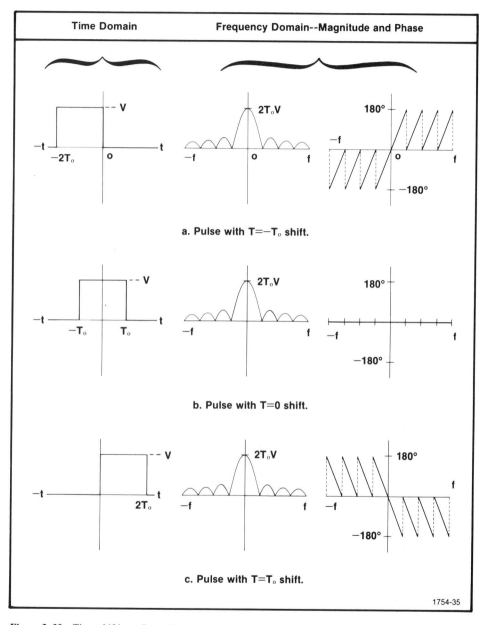

Figure 3–20 Time shifting affects phase only.

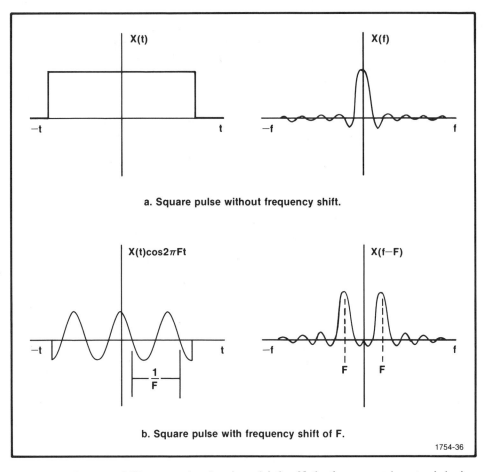

a. Square pulse without frequency shift.

b. Square pulse with frequency shift of F.

1754-36

Figure 3–21   Frequency shifting causes time-domain modulation. Notice the correspondence to windowing in Fig. 3–15.

USING THE FOURIER INTEGRAL— MATTERS OF PRACTICALITY

Though it may be more versatile than the Fourier series, the Fourier integral still faces much the same matters of practicality. Chief among these is the requirement that the waveform be mathematically describable. If you can't write an equation for it, the integral can't be used. Secondary to this is the complexity of the function's description and the mental investment required to transform it. In some cases, it is just not worth doing all the mathematics.

For many common waveforms, you can often obtain Fourier transforms by referring to tables and charts in engineering handbooks and texts. But beyond the scope of standard waveform tables, something else has to be done since real waveforms rarely match the idealized descriptions in tables. This is where waveform digitizing and digital signal processing come in. This is where the discrete Fourier transform (DFT) comes in.

The digital approach to Fourier transformation is the subject of Parts II and III of this book. In these two parts, you will see how the DFT, evaluated by a fast Fourier transform (FFT) algorithm, closely approximates the analog transform. And you will also be introduced to the concepts necessary for practical Fourier analysis. Perhaps more importantly, though, you will see how the FFT can reduce your analysis efforts to little more than what might be considered a standard oscilloscope measurement.

Part II

DIGITAL FOURIER ANALYSIS

The major concepts and properties of the Fourier transform were discussed in Part I. Part I is theory, but theory is frequently difficult to apply directly without some tools. The discrete Fourier transform (DFT) and the fast Fourier transform (FFT) algorithm are your tools for quick and easy application of Fourier theory.

Part II explores digital Fourier analysis and the application of Fourier theory through the DFT. The effects of digitizing a waveform are looked at, and Fourier analysis with the DFT is discussed. The bulk of the discussion, however, centers on the FFT, which is simply an efficient algorithm for computing the DFT.

The major point to keep in mind throughout Part II is that the DFT is viewed as a discrete approximation of the Fourier integral, which is the analog counterpart to the DFT. The quality of the approximation depends upon your understanding of Fourier theory and your ability to apply it to acquiring and digitizing analog signals, to massage the data before transforming it, and to interpret the discrete results. Part II presents the background you'll need for meeting these requirements.

Chapter 4 ——————————————————————

Introduction to the Discrete and Fast Fourier Transforms

The DFT and FFT operate on finite sequences—sets of data with each point discretely and evenly spaced in time. However, the waveforms you usually want to transform—real-life waveforms—are analog in nature. They are continuous in time, and they must be sampled at discrete points before the DFT or the FFT algorithm can be applied. And, to be processed by a digitial computer, these sampled points must be digitized as well. Understanding two basic concepts of the full analog-to-digital conversion, namely windowing and sampling, will put you a long way down the road to appreciating the power of the FFT and understanding its results.

WINDOWING AND SAMPLING ARE OLD IDEAS
WITH NEW NAMES

If you have ever calculated the values of a waveform and plotted them on graph paper, you have done windowing and sampling. To see how these work, let's graph a cosine wave. For example, let's use a cosine wave having a peak amplitude of 1 V (volt) and a frequency of 12.5 Hz. In terms of mathematics, this is represented by $x(t) = \cos 2\pi 12.5t$.

Now, think about graphing $x(t) = 2\pi 12.5t$. When the graph is done, we want everyone to recognize it as a cosine wave. Four repetitions of the waveform should be enough for that, which means we'll need values calculated over a time interval of about $4/12.5 = 0.32$ sec. The waveform should be made obvious without too much interpolation between points, so let's use 32 equally spaced points. In other

1754-37

t	x(t)	t	x(t)	t	x(t)	t	x(t)
0.00	1.000	0.08	1.000	0.16	1.000	0.24	1.000
0.01	0.707	0.09	0.707	0.17	0.707	0.25	0.707
0.02	0.000	0.10	0.000	0.18	0.000	0.26	0.000
0.03	−0.707	0.11	−0.707	0.19	−0.707	0.27	−0.707
0.04	−1.000	0.12	−1.000	0.20	−1.000	0.28	−1.000
0.05	−0.707	0.13	−0.707	0.21	−0.707	0.29	−0.707
0.06	0.000	0.14	0.000	0.22	0.000	0.30	0.000
0.07	0.707	0.15	0.707	0.23	0.707	0.31	0.707

Table 4–1   Values of $x(t) = \cos 2\pi 12.5t$ at every 0.01 second over the interval from 0.00 second through 0.31 second.

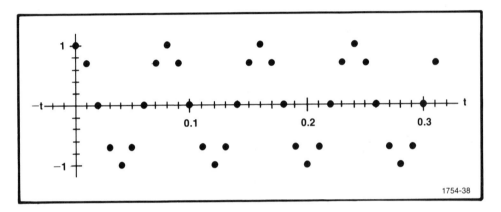

Figure 4–1   A windowed and sampled cosine wave.

words, let's plot the cosine wave at every 0.01 sec from 0.00 sec through 0.31 sec. The values for these points are listed in Table 4–1 and are plotted in Fig. 4–1.

Notice in Table 4–1 that the beginning point of each cycle has a value of 1 and the end point a value of 0.707. The last point value on a cycle is not 1 because that is the beginning value of the next cycle. This pattern is important for later discussion of the FFT, and it will be covered again.

Now take a look at Fig. 4–1, which is a plot of a windowed and sampled cosine wave. It is windowed by virtue of deciding to plot the waveform over a finite time interval, in this case 0.00 through 0.31 sec. It is sampled by virtue of the decision to find and show its actual values at only 32 discrete points in that time interval. So you see, the fundamental concepts of windowing and sampling date back at least to the concept of graph paper.

In terms of waveform processing, Fig. 4–1 can be thought of as being obtained through the multiplication shown in Fig. 4–2. Figure 4–2a is the waveform, in this case a cosine wave, that is going to be windowed. Figure 4–2b is the rectangular window (square pulse) that is going to be used. Their product, shown in Fig. 4–2c, is a rectangularly windowed cosine wave. In Fig. 4–2b and c, notice the heavy dots

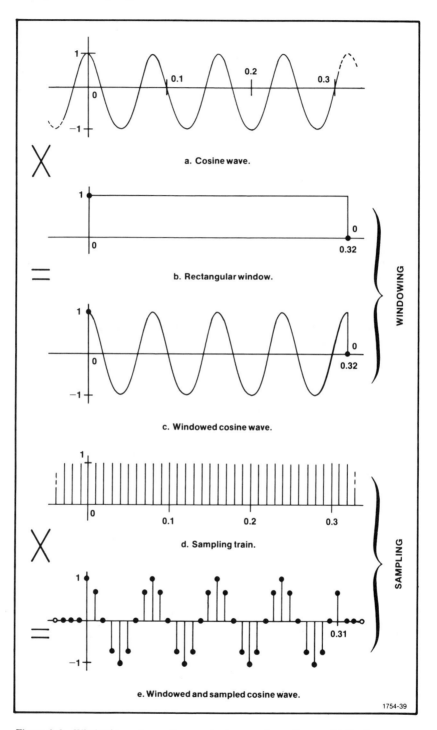

a. Cosine wave.

b. Rectangular window.

c. Windowed cosine wave.

WINDOWING

d. Sampling train.

e. Windowed and sampled cosine wave.

SAMPLING

1754-39

Figure 4–2   Windowing and sampling equates to several waveform multiplications.

at the window edges. These aren't actually part of the window, but are illustrative devices for signifying the values at the window edges. The window in Fig. 4–2b is defined to begin at time zero with an amplitude of 1 and continues up to 0.32 sec with a constant amplitude of 1. Then the window ends at 0.32 sec with an amplitude of zero. Such precise definition of window edges becomes more important in later discussions of the periodicity assumed by both the DFT and the FFT.

After windowing, sampling is represented by the train of unity amplitude impulses shown in Fig. 4–2d. Each impulse in this train rises from zero and returns to zero, all instantaneously. Thus, each impulse is nonzero at only a single, discrete time. Also, each impulse encloses an area equal to 1, and the train of impulses is graphically represented by a train of spikes, each having a height equal to its area. When this impulse train, or sampling train, is multiplied with the windowed cosine wave, the product (Fig. 4–2e) is a windowed and sampled cosine wave. Each impulse then has an area corresponding to the cosine value at that point. The time locations of the samples correspond to the time locations of the impulses. And, since each impulse exists at a single, discrete time point, each sample exists at a single, discrete time point. The discrete sample times and the sample values are indicated in Fig. 4–2e by the heavy dots.

Notice that the windowed and sampled cosine wave in Fig. 4–2e is exactly the same as the graphed cosine wave in Fig. 4–1, except the impulses aren't shown in Fig. 4–1. Also, the values in Table 4–1 apply equally to Figs. 4–1 and 4–2e. So Fig. 4–2 is just another way to get the results of Table 4–1 and Fig. 4–1.

In terms of physical processes, the results of Fig. 4–2e and Table 4–1 can be obtained by using an analog-to-digital converter. This does not actually involve physical generation and multiplication of waveforms as shown in Fig. 4–2, but the conversion is analogous and produces the same end result. For example, the window in Fig. 4–2b corresponds to triggering the input of an analog-to-digital converter on and off. When the converter is triggered on, the waveform to be digitized is allowed to pass. When the converter is switched off, the waveform is blocked. What passes is equivalent to multiplying the waveform by a square pulse. The window length or pulse width corresponds to the length of time the analog-to-digital converter is on.

In practice, sampling is done by the same means. The windowed waveform is looked at through much smaller, closely spaced sampling windows. When the sampling window is gated on, a capacitor or some other device is allowed to assume the value of the waveform during the sample time. A single value is thus obtained at each sample point, as shown in Figs. 4–1 and 4–2e. These values can then be stored in a memory corresponding to Table 4–1. In most cases, though, the sample values are converted to digital data and stored in a memory compatible with a computer. Then the computer or some related digital device can be used to process the data.

There are a number of ways that windowing and sampling can be done. As was just discussed, some analog-to-digital converters window first and then sample. Others sample first, then window the desired block of sampled data. In still others, windowing and sampling occur at the same time. In any case, you'll find that the end result is always the same—a windowed and sampled waveform.

There are, however, different windowing shapes and sampling schemes that may be used. Figure 4–2 uses a square pulse for a window. A cosine-squared pulse could have been used instead, and the results would have been different. These various window shapes and the results of their use will be covered in a later chapter.

As far as sampling is concerned, samples can be taken over equally spaced intervals, logarithmically spaced intervals, or by any other spacing scheme. But the use of equally spaced samples is probably more widespread than any other technique, and further discussion here is confined to cases using equal-spaced sampling. When the term *sampling* is used, it should be taken to mean "equal-spaced sampling" unless otherwise specified.

THE DFT WORKS ON SAMPLED DATA

Figure 4–2e can be transformed to the frequency domain by applying the Fourier integral over the window interval to the product of the waveform and impulse train. In fact, if you sit down with pencil and paper and do the transform, paying attention to notation and performing the correct manipulations, you'll come up with an expression known as the *discrete Fourier transform* (*DFT*), as follows:

$$X_d(k\Delta f) = \Delta t \sum_{n=0}^{N-1} x(n\Delta t)e^{-j2\pi k\Delta fn\Delta t}$$

This expression allows you to transform a time series of samples, such as in Table 4–1, to a series of frequency-domain samples. By some additional manipulations, you can also develop the inverse DFT, which is

$$x(n\Delta t) = \Delta f \sum_{k=0}^{N-1} X_d(k\Delta f)e^{j2\pi k\Delta fn\Delta t}$$

This expression allows you to transform a series of frequency-domain samples, computed by the DFT, back to a series of time-domain samples.

In both of these discrete expressions, the variables have the following definitions:

N = number of samples being considered.

Δt = the time between samples, referred to as the *sampling interval*. From this, $N\Delta t$ gives the window length, often referred to as the *time record length*.

Δf = the sample interval in the frequency domain and is equal to $1/N\Delta t$.

n = the time sample index. Its values are $n = 0, 1, 2, \ldots, N - 1$.

k = the index for the computed set of discrete frequency components. Its values are $k = 0, 1, 2, \ldots, N - 1$.

$x(n\Delta t)$ = the discrete set of time samples that defines the waveform to be transformed.

$X(k\Delta f)$ = the set of Fourier coefficients obtained by the DFT of $x(n\Delta t)$.

e = the base of the natural logarithm.

j = the symbol of complex notation, indicating the imaginary part of a complex quantity ($j = \sqrt{-1}$).

By continuing with some substitutions, letting $\Delta t = 1$ so that $\Delta f = 1/N$, you can arrive at the more common form of the DFT and its inverse as follows:

$$\text{DFT:} \qquad X_d(k) = \sum_{n=0}^{N-1} x(n)e^{-j2\pi kn/N}$$

$$\text{inverse DFT:} \qquad x(n) = \frac{1}{N} \sum_{k=0}^{N-1} X_d(k)e^{j2\pi kn/N}$$

Since the $1/N$ before the summation in $x(n)$ is simply a scaling term, it can be included in either—not both—the direct or inverse expression. For the examples given here, $1/N$ is shifted to the direct expression, $X_d(k)$. This alternate formulation gives the DC term, $X_d(0)$, as the arithmetic mean of the time samples.

For computational convenience, *Euler's identity* ($e^{\pm j\theta} = \cos\theta \pm j\sin\theta$) is used to change the complex exponential to give the DFT and inverse DFT as follows:

$$X_d(k) = \frac{1}{N} \sum_{n=0}^{N-1} x(n)\frac{\cos 2\pi kn}{N} - jx(n)\frac{\sin 2\pi kn}{N}$$

and

$$x(n) = \sum_{k=0}^{N-1} X_d(k)\frac{\cos 2\pi kn}{N} + jX_d(k)\frac{\sin 2\pi kn}{N}$$

Having developed this set of expressions, it is now a fairly straightforward set of operations to compute the discrete Fourier transform for any string of waveform samples. For example, you can compute the DFT of the cosine wave from the values listed in Table 4–1. There are 32 samples given there, so k and n take on values of 0, 1, 2, . . . , 31. The values of $x(n)$ are taken directly from the table, where $x(0)$ corresponds to time zero and has a value of 1.000, $x(1)$ corresponds to 0.01 sec and has a value of 0.707, and so on. Each Fourier coefficient, $X_d(k)$, is computed by summing $[x(n)\cos 2\pi kn/N - j\sin 2\pi kn/N]$ for all values of n at each k. For example, let $k = 0$ and sum $[x(n)\cos 2\pi kn/N - j\sin 2\pi kn/N]$ for $n = 0$ to 31. Then let $k = 1$ and do the summation again for all n. This goes on until the summation for $k = 31$ is reached and completed. When that is done, you have the set of 32 Fourier coefficients for the 32 time-domain samples of the cosine wave. These Fourier coefficients define the cosine wave's complex frequency domain at 32 discrete frequencies.

If you've taken up pencil and paper to try the DFT on Table 4–1, it won't take you long to realize that there are 32 terms to be summed for each of the 32 values of k. That means there are $32 \times 32 = 1024$ major operations for a 32-point DFT. Although each operation is in itself fairly easy, doing 1024 of them is not. There has to be a better way!

A DFT Program Makes It Easier. Computing the DFT is essentially a repetitive task. The major operations are the same over and over again. What changes are the index values, k and n, and the sample values, $x(n)$. So the DFT is well suited to evaluation by computer program.

Put away your pencil and paper. Figure 4–3 lists a BASIC program that generates the values in Table 4–1, computes the DFT, and outputs the results in tabular form.

If you have a computer with BASIC software handy, enter the program from Fig. 4–3 and run it. You may have to change some syntax to match your software, particularly when it comes to the PRINT statements in lines 300 through 320. The main thing, though, is just to print out the results by index values (line 315). Once you have the program running, your output should match the left half of Table 4–2.

There are a lot of numbers in the left half of Table 4–2. But notice that they are all very small compared with the two numbers shown in bold print (0.5 and 0.5002). The smaller numbers are digital noise and can be considered to be zero compared to the two major values, 0.5 and 0.5002, which are the spectral components of the cosine wave given in Fig. 4–1.

Next, notice below the table of Fourier coefficients that a spectral plot has been made in the order of index values. At first there may be some confusion over this plot. But, with a little study, it can be discerned that the coefficients for indices 16 through 31 are for the negative frequency domain and are in inverse order of frequency as well. To set things right, the right half of the plot should be picked up and placed on the left side of the DC point.

From the foregoing, it's apparent that the DFT doesn't necessarily yield the coefficients in a "normal" order. So some extra care (or additional programming steps) is needed to keep track of which indices represent which frequencies. Most

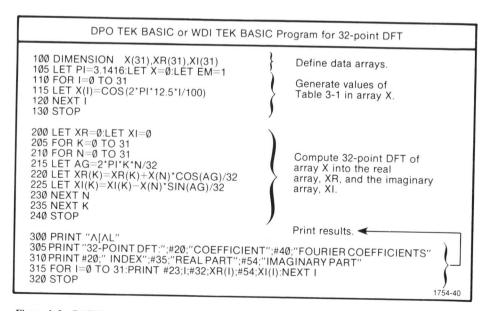

Figure 4–3 BASIC program for computing a 32-point DFT.

Index	DFT Real Part	DFT Imaginary Part	FFT Real Part	FFT Imaginary Part
0	−3.052E-5	0	0	0
1	0	−7.629E-6	3.052E-5	0
2	−3.052E-5	−7.629E-6	3.052E-5	0
3	0	2.289E-5	3.052E-5	0
4	**0.5**	−2.289E-5	0	3.052E-5
5	6.104E-5	−3.815E-5	−3.052E-5	6.104E-5
6	−3.052E-5	−7.629E-6	0	6.104E-5
7	−3.052E-5	−3.815E-5	0	6.104E-5
8	0	−3.052E-5	0	0
9	1.221E-4	−6.104E-5	0	0
10	−6.104E-5	1.526E-5	0	0
11	0	3.052E-5	3.052E-5	0
12	3.052E-5	0	**0.5**	−3.052E-5
13	6.104E-5	−3.815E-5	0	−3.052E-5
14	0	−6.867E-5	3.052E-5	−3.052E-5
15	−3.052E-5	6.867E-5	3.052E-5	−3.052E-5
16	0	1.526E-5	3.052E-5	0
17	0	4.578E-5	3.052E-5	3.052E-5
18	3.052E-5	−4.578E-5	3.052E-5	3.052E-5
19	3.052E-5	2.289E-5	0	3.052E-5
20	3.052E-5	−9.918E-5	**0.5**	0
21	6.104E-5	5.341E-5	3.052E-5	0
22	−1.631E-4	−9.918E-5	0	0
23	−1.221E-4	8.392E-5	0	0
24	−1.221E-4	3.815E-5	0	0
25	−6.104E-5	9.155E-5	0	−6.104E-5
26	−1.526E-4	1.526E-5	0	−6.104E-5
27	6.104E-5	4.578E-5	−3.052E-5	−6.104E-5
28	**0.5002**	3.052E-5	0	−3.052E-5
29	1.221E-4	−1.144E-4	3.052E-5	0
30	4.273E-4	1.144E-4	3.052E-5	0
31	−2.136E-4	−1.068E-4	3.052E-5	0

DFT annotations: DC (index 0); positive frequency (indices 1–15); Nyquist frequency (index 16); negative frequency (indices 17–31).

FFT annotations: Nyquist frequency (index 0); negative frequency (indices 1–15); DC (index 16); positive frequency (indices 17–31).

DFT plot: n=0 ·4 ─ ─ ─ 16 ─ ─ ─ 28·31; 0.5; DC; +12.5 Hz; −12.5 Hz; *Nyquist frequency

FFT plot: n=0 ─ ─ ─ 12·16·20 ─ ─ ─ 31; 0.5; DC; −12.5 Hz; *Nyquist frequency; +12.5 Hz

1754-41

Frequency interval is given by $\Delta f = 1/N\Delta t$. Bold type indicates frequency components for cosine wave in Table 4-1; others are effectively zero.

*The Nyquist frequency is the highest frequency sinusoid that can be defined at a given sampling rate. For equally spaced samples (Δt), the Nyquist frequency is $1/2\Delta t$.

Table 4–2   32-point Fourier transformation of the data in Table 4–1.

programs, especially commercial programs, reformat the output results for convenient order of tabulation or viewing on a screen. The exact method of display depends on the program's output routine and is generally described in the program's documentation.

Taking a moment here for digression, it's also interesting to note that there were four cycles of cosine wave in the window operated on by the DFT and that the computed spectral components each occurred at the fourth indices from the DC point.

Four cycles, fourth index values from DC—is this a coincidence? No. Running the program again for eight cycles in the window (25-Hz cosine wave) would result in the spectral component occurring at the eighth point from DC. Three cycles would have landed on the third point, seven cycles on the seventh. There is a relationship, and it will be further explored later on.

For now, however, there's another point to be made from the DFT program in Fig. 4–3. If you were able to enter the program and run it on a computer, you may have discovered that it took more than a blink of an eye to execute. Or maybe you used a very fast machine and didn't notice. If so, modify the program to do a 512-point DFT instead of a 32-point DFT. That's $512 \times 512 = 262,144$ major operations for evaluation. You'll have enough time now to go out for a cup of coffee before the program completes, even if you have a very fast computer.

Then, if your software has an FFT algorithm, try using it. Its syntax is probably something like FFT X,XR,XI, where X is the array of 512 data points that you want to transform and XR and XI are arrays for holding the real and imaginary parts of the results. You'll probably find that the FFT algorithm will complete the 512-point transform before you can even take a sip of your coffee!

Why the difference? Well, first of all, your FFT algorithm may have only computed half of the Fourier coefficients, those for the positive frequency domain. There's no sense computing the negative frequency domain, since it is a duplicate of the positive half for real functions of time. That shortcut saves some time. But there are some bigger differences. For example, the DFT program in Fig. 4–3 is written in a high-level language (BASIC). It would execute faster if it were implemented in a low-level assembly language, in which most FFT algorithms are written. But even then, the direct evaluation wouldn't be as fast as the FFT algorithm. The FFT algorithm is, in short, the most elegant and time-efficient way to compute Fourier coefficients.

At this point, there is often a tendency to attach too much significance to the FFT. True, its speed is significant. But the FFT is not really anything different from the DFT. It is just an algorithm for computing the DFT. Exactly how it works is not as important in most cases as remembering the concepts of Fourier theory and how they lead up to the FFT through the DFT.

Look again at Table 4–2. Except for some digital noise and the order of the coefficients, the FFT results and DFT results are the same. That's why the FFT was developed, to get the same results faster. Perhaps there's no better way to reiterate this than to look at some of the history surrounding the FFT algorithm.

HOW THE FFT ALGORITHM CAME ABOUT

When you consider the 32^2 major operations required for the 32-point DFT in the preceding example, it isn't difficult to understand why discrete Fourier analysis was generally avoided by scientists working before the development of the modern digital computer. And, for many applications, 32 samples are not enough. The required number, N, of samples for defining many real-life functions frequently runs into the hundreds, sometimes a thousand or more. So the prospect of N^2 operations by hand calculation was enough in the past to discourage discrete Fourier techniques as an analysis tool.

But some fields of study beg for Fourier analysis. In fact, some information can only be gained by Fourier analysis, and there is nothing to do but plunge forward with the calculations.

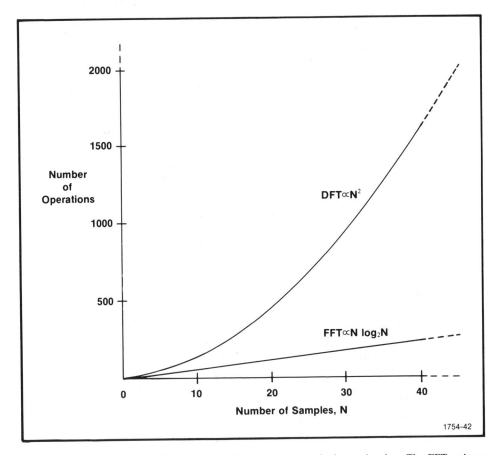

Figure 4-4 How fast is fast? The more operations to be done, the longer it takes. The FFT reduces the number of operations, and for large N, the time advantage is tremendous. For example, the FFT for 512 samples has a better than 50 to 1 advantage over the DFT.

Accordingly, it was standard in the days of hand calculation to be as concerned with minimizing as well as application possibilities. Scientists welcomed papers describing means for reducing the number of calculations in an analysis technique almost as much as they welcomed the new analysis technique itself. It was in such a minimization paper (for 12- and 24-point transforms) published in 1903 that C. Runge described the technique that later became known as the *FFT*. Later, in 1942, a more generalized approach was advanced by Danielson and Lanczos. By recognizing certain symmetries and periodicities, they reduced evaluation of the DFT for $N = 2^k$ points to $N \log_2 N$ major operations. The significant savings of $N \log_2 N$ operations versus N^2 operations is apparent in Fig. 4–4.

Even with the savings indicated in Fig. 4–4, hand evaluation is still a staggering task for reasonable values of N. For example, using an efficient algorithm, L. H. Thomas of the IBM Watson Laboratory reported spending three months in 1948 doing a transform with the aid of an office accounting machine. It's no wonder people avoided Fourier analysis whenever possible!

Then came the digital computer, but, without widespread use, the techniques of Danielson and Lanczos were generally unknown. As a result, computer evaluation was done by the direct approach using N^2 operations. Even with a computer, N^2 operations took too much time for large N. So the cost of doing Fourier analysis by discrete methods was prohibitive in many cases.

Still, Fourier analysis was necessary to some research. In the early 1960s, Richard L. Garwin was studying solid helium and had a great need for Fourier techniques. His need caused him to contact a colleague, James W. Tukey, and ask for an efficient means of evaluating the discrete Fourier transform. Tukey supplied him with the essence of what was to become known as the fast Fourier transform. Then Garwin approached the director of mathematical sciences at IBM with the problem of programming the algorithm. As a result, J. W. Cooley became involved. In Cooley's words at his keynote address at the 1968 Arden House Workshop on Fast Fourier Transform Processing,

> Garwin came to the computing center at IBM Research in Yorktown Heights to have the algorithm programmed. I was new at the computing center and was doing some of my own research. Since I was the only one with nothing important to do, they gave me this problem to work on. It looked interesting, but I thought that what I was doing was more important; however, with a little prodding from Garwin, I got the problem out in my spare time and gave it to him. It was his problem and I thought I would hear no more about it and went back to doing some real work.*

But Cooley hadn't heard the last of it. Garwin saw a wide range of applications for the program besides studying helium and began contacting various scientists and making the possibilities known. "As a result of his publicity," Cooley relates on the same occasion, "I started to get letters requesting programs and write-ups. Requests

---

* *IEEE Transactions on Audio and Electroacoustics*, vol. AU-17, no. 2, June 1969.

for a paper then started arriving. I was asked to write a paper and did so, asking Tukey to co-author it. He did, and the paper in *Mathematics of Computation*, in 1965, was the result." This paper describes the Cooley–Tukey algorithm for evaluating the discrete Fourier transform.

For the most part, the Cooley–Tukey algorithm is simply referred to as the *FFT*. There are also a number of other algorithms, offshoots of the Cooley–Tukey algorithm, that are lumped under the term *FFT*, and any algorithm that provides the $N \log_2 N$ advantage is generally referred to as an FFT.

PUTTING THE ALGORITHM TO WORK— HARDWARE, FIRMWARE, OR SOFTWARE?

Garwin's approach to the FFT was a software approach. General-purpose computers were available to him, and he had already obtained by computer simulations a very large amount of nuclear spin data for helium. All he needed was a program to Fourier transform the data. A software implementation of the FFT was a natural approach.

Software implementation of the FFT has some distinct advantages too. The FFT algorithm can be programmed in a general enough manner so that the number of points to be transformed can vary over a tremendous range. If you have 512 samples, the program does a 512-point FFT; if you have 1024 or 4096 samples, the same approach is taken. With the flexibility offered by software, the FFT algorithm can do any length transform you desire, within the limits of the specific routine and computer memory size. (Quite often, though, the length is restricted to a power of two—2, 4, 8, 16, 32, ... This offers algorithm simplicity and additional execution speed.) Also, with general-purpose software, you can choose to do a variety of further processing or analysis if you like, and you can put the results in just about any form you want—tables, graphs, diagrams, and so forth. In short, you make the software FFT fit the data instead of making the data fit the FFT.

Generally, what has been said for software also applies to firmware implementations of the FFT. With recent advances in firmware components and firmware design, FFT algorithm implementation can be given all of the extent and flexibility of software implementations. The advantage of firmware is that it is generally faster executing. However, the disadvantage is that its portability depends upon hardware fixturing and compatibility.

There are some trade-offs in a general-purpose software or firmware implementation, though. The options that make it general purpose require decisions and interpretations by software, and this takes additional execution time. Of course, execution can be speeded by tailoring the FFT routine to a specific need, but then the program may not be flexible enough to handle other applications.

Another trade-off is that a software FFT is, strictly speaking, an off-line operation; the signal data is acquired, stored, and then processed. The same is true for firmware. In both cases, the results are not real-time results, but occur some finite time after the input signal. However, from a practical viewpoint, real-time results often don't have to be instantaneous results. If the time between acquiring a signal and getting FFT results is milliseconds and signal variations are on the order of seconds or more, then the FFT results might effectively be considered to be real-time results. On the other hand, signal variations on the order of microseconds would disqualify a millisecond return of results from being considered as real-time results. But then, a large block of FFT applications do not require real-time results.

As a final point, a software implementation of the FFT algorithm simply does not execute as fast as the corresponding hardware implementation. Compared with a software approach, where a general-purpose machine is directed in every step of the task, a hardware FFT has the instructions built into it. It just goes ahead and does its specific task—no questions asked and usually very little flexibility offered. In fact, a dedicated hardware FFT processor designed for a specific, narrow application can be so fast as to be considered an on-line analyzer, providing results that are real-time for all intents and purposes. But you can't tell it to do anything but that one specific task. You have to make your data fit the hardware FFT.

Regardless of the FFT implementation, the final results are the same for the same algorithm. The algorithm itself is simply a method of evaluation. How you choose to implement the method—hardware, firmware, software, or pencil and paper—makes no difference in the results. The hardware FFT of a waveform is the same as the firmware or software FFT of the same waveform, and these are the same as the "pencil-and-paper" FFT for that same waveform. The major consideration is how quickly you get the results. And this depends on algorithm design and the computer's instruction speed, or how fast you are with a pencil and paper.

Chapter 5

Understanding FFT Results—
Basic Data Formats

A BRIEF RECAP AND A LOOK AHEAD

In Chapter 2, continuous periodic signals and their transformation to the frequency domain by the Fourier series were discussed. Then, in the following chapter, continuous, nonperiodic signals and their transformation by the Fourier integral were covered. Accompanying that, the idea of transforming windowed signals with the Fourier integral and the concept of interpreting results from a periodic or nonperiodic point of view were introduced.

Then, in Chapter 4, another step was taken. Windowed signals were sampled and transformed by the discrete Fourier transform. And, finally, the FFT was introduced as an efficient method of evaluating the DFT.

The DFT, or FFT, is a third type of Fourier transform. It is related to the Fourier series and Fourier integral, but it is defined only for discrete values over a finite interval. From a digital viewpoint, the FFT provides the exact transform for the discrete values provided. Either a continuous, periodic signal or a continuous, nonperiodic signal can be windowed and sampled to provide discrete values. These discrete values can then be transformed to the frequency domain by the FFT. The result is exactly as it should be when viewed from the digital perspective as only a representative of the analog world. That is, the FFT provides the correct frequency-domain information for the windowed and sampled version of the waveform. If, however, you wish to interpret the FFT results from a continuous viewpoint, as is usually the case, then the interpretation must be done with care. An interpretation from the continuous, analog viewpoint must always be tempered with the realities of the windowed and sampled data.

This chapter looks at FFT results in general. Chapter 6 goes on to explore the effects of windowing and sampling and how they influence FFT results from a continuous viewpoint.

For convenience, the FFT examples used in these chapters use the Sande–Tukey FFT algorithm as implemented by Tektronix, Inc., in their DPO TEK BASIC and WDI TEK BASIC software. The same algorithm also appears in their later signal processing packages and ROM packs, but modified to present only the positive frequency domain. (It should be noted that, where the original Cooley–Tukey algorithm uses a process called *decimation in time*, the Sande–Tukey algorithm uses *decimation in frequency*. While the methods differ, the two algorithms produce the same final results. It's just a different path.)

Also, because of their convenient size, displayed FFT results are provided as photos taken from the cathode ray tube of a Tektronix, Inc., digitizing oscilloscope. But, although these examples are obtained via specific software and instrumentation, keep in mind that the concepts illustrated are applicable to FFTs in general.

And finally, the examples in this and subsequent chapters deal with 512-element arrays. However, arrays of other lengths can, in practice, be used with the FFT. For example, a 32-element array was used in Chapter 4 to introduce the DFT and FFT.

PROVIDING TIME-DOMAIN DATA FOR THE FFT

In the most fundamental sense, there are two types of time-domain information available for any type of analysis. There is information that is directly measurable or observable. Then there is theoretical information generated by mathematical formulas or other means of speculation.

For the case of measurable information, like the waveform in Fig. 1–10 of Chapter 1 that was used to point out the need for Fourier analysis, the information can be acquired and sampled by standard measurement techniques. Then the samples from the acquisition window can be digitized and stored in a data array for further processing. In the case of the software used for the examples here, the length of the standard waveform data array is 512 elements. Each array element corresponds to a waveform sample from the acquisition window. Array element zero corresponds to time zero at the left edge of the window, as indicated in Fig. 5–1, and array element 511 corresponds to one sample less than the maximum time at the right edge of the window. The remaining elements, 1 through 510, correspond to equally spaced points between the window edges.

The digital values stored in each array element correspond to the sampled values of the waveform at equally spaced time points. In short, the waveform array corresponds to the table of values you might construct from point-by-point measurement of the waveform's amplitude. The sampled waveform shown in Fig. 5–1 would then

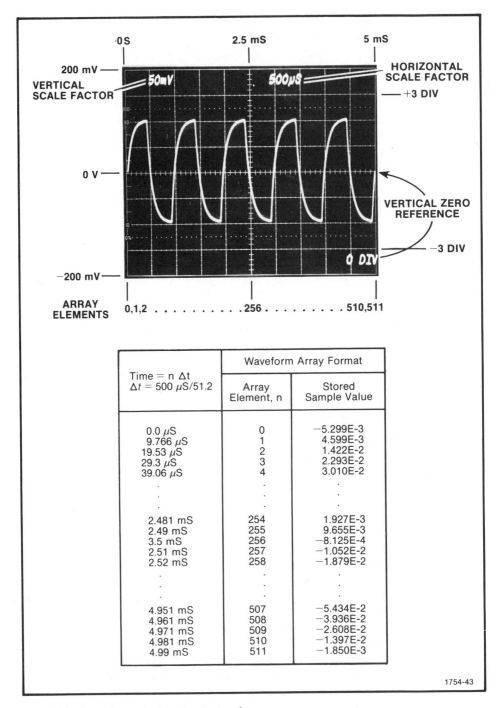

Figure 5–1   Array format for time-domain waveforms.

correspond to a graph you might draw from the table of point-by-point measurements.

For the case of theoretical data, the computed values are simply entered into the data array in much the same manner that Table 4–1 in Chapter 4 was constructed. However, in the case of Table 4–1, the array would be a 32-element array. To place the same theoretical cosine wave over the same window into a 512-element array, you would need to compute samples at every 0.626 msec (0.32 sec divided by 511 time increments) from 0 sec to 0.32 sec. In either case, an FFT of the theoretical data can be done. In the former, a 32-point FFT is done, and a 512-point FFT is done for the latter. The major difference, of course, is that 512 samples over the same time interval provide much better time resolution and result in calculation of more frequency-domain points.

The great majority of measured and theoretical time-domain data is real-valued data. That is, the data is real, not necessarily in the sense that they are physical, but real in the sense that they are not complex-valued data. They do not have an imaginary part, or at least the imaginary part isn't considered. Thus, you only need one waveform array to store real time-domain data. However, two additional waveform arrays are required for storage of the corresponding frequency-domain data. One of these arrays is for the real part or the magnitude, and the other is for the imaginary part or phase, depending on whether the results are presented in rectangular or polar form.

Additionally, you'll need to be aware of the size requirement for these arrays. If your software provides FFT results for just positive frequencies, each array for results needs only be one more element longer than half the length of the time-domain array. For example, a 512-element time-domain array requires two frequency-domain arrays of 257 elements each to provide room for the Fourier coefficients from DC (left edge of display) out to the Nyquist frequency (F_N = half the sampling frequency). Or, if your FFT algorithm provides results for both the negative and positive frequency domain, the two frequency-domain arrays need to be equal in length to the time-domain array. For this latter case, DC will be at one of the center array elements (either 255 or 256, depending on whether the Nyquist frequency is displayed at the positive or negative frequency extreme).

These are all general rules that apply in most cases. However, before using any implementation of the FFT, you should check its accompanying documentation for specific details on its array-size requirements and how the output is formatted.

FFT RESULTS ARE NORMALLY IN RECTANGULAR FORM

Looking again at the mathematics of the DFT (which the FFT computes in a more efficient manner),

$$X_d(k) = \frac{1}{N} \sum_{n=0}^{N-1} x(n) e^{-j2\pi kn/N}$$

or more conveniently as

$$X_d(k) = \frac{1}{N} \sum_{n=0}^{N-1} x(n) \left(\cos \frac{2\pi kn}{N} - j\sin \frac{2\pi kn}{N} \right)$$

it is apparent that the results come in two parts—a real part and an imaginary part preceded by the imaginary operator, j. This is the rectangular form of results provided by most FFT implementations.

As a specific example, let's look at transforming the time-domain waveform in Fig. 5–1 to the frequency domain. The FFT implementation used for this example is that of **DPO TEK BASIC**, a signal processing software package developed by Tektronix, Inc. Again, the point here is not to dwell on a particular algorithm or implementation, but to simply look at one way of formatting FFT results.

Proceeding with the example, let's say that the waveform to be transformed is stored in the 512-element array designated as array A. A display of this data is shown in Fig. 5–2a. This array of waveform data, array A, is transformed to the frequency domain by the statement

FFT A,B,C

where B and C are also 512-element arrays for storing the frequency-domain results, which, in this case, span both positive and negative frequencies. The real part of the frequency domain is stored in array B (and is displayed in Fig. 5–2b); the imaginary part is stored in array C (and is displayed in Fig. 5–2c). Notice how these results closely resemble the line spectrum of a periodic waveform. A periodic point of view follows naturally from Fig. 5–2a.

The arrangement of the Fourier coefficients in the two frequency-domain arrays is such that 0 Hz (DC component) occurs at array element 256. This corresponds to the center of each display in Fig. 5–2b and c. The positive portion of the frequency domain extends to the right of center (elements 257 through 511), and the negative frequency domain extends over the left half of the display (elements 0 to 255). For the specific software used, element 0 (left edge of display) corresponds to the highest frequency that can be defined by the time-domain sample interval and is referred to as the *Nyquist frequency*.

Other software packages may format the results in a slightly different manner. For example, the Nyquist frequency can appear as the last element in the array (corresponding to the right edge of the displayed data). This is usually the case when the FFT implementation presents just the positive frequency domain. Also, for a display of positive frequencies only, the left edge of the display will correspond to DC or 0 Hz.

Whatever the format of the results for a given FFT implementation, the same format is generally required for frequency data to be inverse Fourier transformed to the time domain. The inverse Fourier transform (IFT) routine uses essentially the same FFT algorithm with some minor changes to the input and output routines. So, in the case of the example in Fig. 5–2, the frequency-domain arrays (B and C)

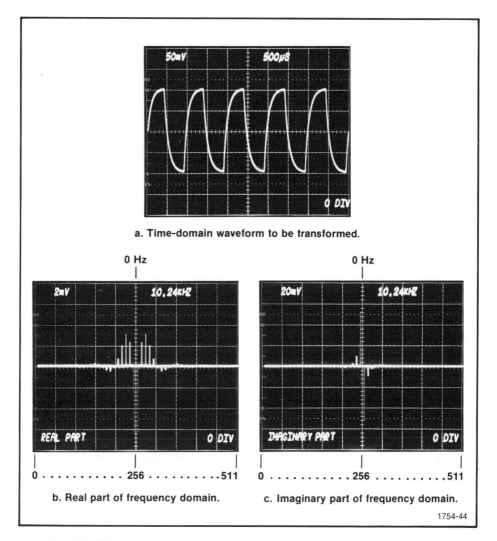

a. Time-domain waveform to be transformed.

b. Real part of frequency domain.

c. Imaginary part of frequency domain.

1754-44

Figure 5–2   The FFT of a time-domain waveform results in two arrays that represent the complex frequency domain in rectangular form.

could be inverse transformed back to another time-domain array of data (D). The result of this inverse transform (using a statement similar to IFT B,C,D) is shown in Fig. 5–3a. The difference between this result from the IFT and the original waveform (Fig. 5–2a) is shown in Fig. 5–3b. Notice that this difference is slight—effectively zero compared with the waveform magnitudes—and can be attributed to arithmetic roundoff error.

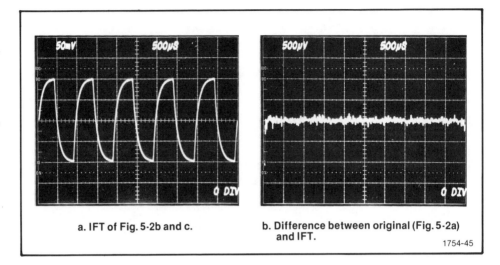

| a. IFT of Fig. 5-2b and c. | b. Difference between original (Fig. 5-2a) and IFT. |

1754-45

Figure 5–3   The IFT of the frequency domain in rectangular form returns the original, time-domain waveform.

FFT RESULTS CAN BE IN POLAR FORM

The *rectangular form* is the most direct result for the FFT and is also more appropriate for further processing. But most people are not entirely comfortable viewing FFT results in rectangular form. Maybe it's because their first introduction to Fourier theory and the frequency domain is through magnitude and phase, the *polar form*. Widespread use of spectrum analyzers may also account for this seeming preference of the polar form. They present frequency-domain information as magnitude data (spectrum analyzers using analog techniques do not, however, provide phase data). Whatever the reasons, people just seem more comfortable viewing FFT results in polar form. And, if you're simply interested in the magnitude of a specific frequency, then the polar form is naturally desirable.

Converting FFT results from rectangular to polar form is a simple operation. Magnitude is obtained from the real and imaginary parts by taking the square root of the sum of the squares (magnitude $= \sqrt{Re + Im}$). Phase is the arctangent of the imaginary part divided by the real part (phase $=$ arctan Im/Re).

Modulo 2π Phase.   For convenience, some FFT implementations contain a polar argument or switch. This extra argument lets you specify results in polar form. In other words, a separate polar conversion routine or command is provided.

For the polar switch situation, the FFT algorithm isn't changed by adding the argument. The intermediate results are still in rectangular form. The polar switch simply causes an additional computational step to convert the results to polar form. Additionally, the format of the result arrays remains the same, but one array contains magnitude and the other, phase. The magnitude array is the result of computing the square root of the sum of the squares of the real and imaginary parts, and the phase array is the result of computing the arctangent of the ratio of the imaginary to real parts.

With the polar conversion command, the process is the same. It is just an additional program command following the FFT command. Or, if neither a polar switch nor a polar conversion command is provided, you can achieve the same results with programming steps to compute the square root of the sum of the squares for magnitude and the arctangent for phase.

In all cases, an arctangent routine is required in the software. It should also be noted that most arctangent routines return results in the range of 2π radians, running from $-\pi$ to π. Thus, the phase resulting from using a polar conversion will be in that range and is often referred to as *modulo 2π phase*.

An example of FFT polar results is shown in Fig. 5–4. The time-domain waveform in Fig. 5–4a is the same waveform used for the rectangular example. The polar form results of the FFTed waveform are shown in Fig. 5–4b and c. The magnitude portion is fairly clear and is what would be expected for a low-pass filtered square wave, which it is. The magnitude spectrum also closely resembles a line spectrum; thus, it is appealing here to use the continuous, periodic point of view. However, the phase portion in Fig. 5–4c does need some further explanation.

First of all, phase can't exist for a frequency component that doesn't exist. Yet, if we view Fig. 5–4a as a periodic waveform and interpret the FFT magnitude results in Fig. 5–4b as a line spectrum, Fig. 5–4c seems to be contradictory. There appears to be phase in Fig. 5–4c at frequencies where the magnitude in Fig. 5–4b appears to be zero, that is, at frequencies between the spectral lines. In reality, though, the magnitude spectrum in Fig. 5–4b is not discrete in the sense of the strictly periodic point of view. Also, the real and imaginary parts back in Fig. 5–2b and c, from which the magnitude and phase were computed, are not line spectra either. All of these spectra have low-level "noise" components residing between the higher level harmonics. For interpreting results from a continuous, periodic point of view, the low-level noise components are low enough to be considered zero. However, from a computing standpoint, these low-level components still exist and are significant in terms of computing phase. For example, a value of 10^{-6} might, for the purpose of interpretation, be considered zero when compared with a value of 1. But, for the purpose of computer processing, arctangent 1/1 and arctangent $10^{-6}/10^{-6}$ are the same—they both are equal to 45° (0.7854 radians).

So even the smallest components can contribute significantly to phase computations. This is desirable when you are looking for information contained in low-level components, but for the case of the present example, it may lead the uninformed to confusion.

How do low-level components come about? Well, in reality, Fig. 5–4a doesn't show an ideal, continuous, periodic waveform. The waveform used for this example was an actual, physical waveform. It was subject to distortions, noise, and all of the other vagaries of reality. So some of the low-level components were actually

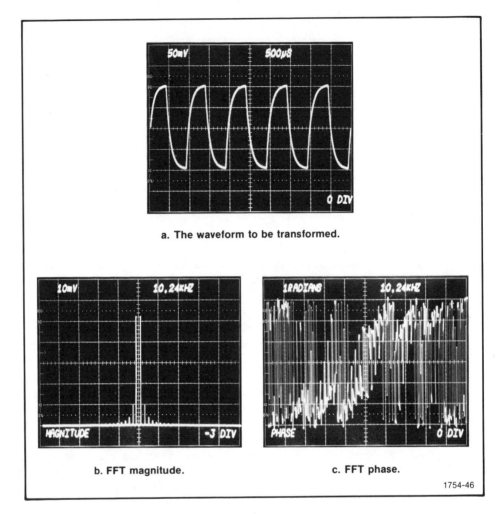

a. The waveform to be transformed.

b. FFT magnitude.

c. FFT phase.

1754-46

Figure 5-4 FFT results can be expressed in polar form.

present in the waveform as broadband noise. Adding to this, the waveform was windowed and sampled, and the samples were digitized—more opportunity for low-level noise to enter the picture. Then the digitized samples were transformed by an FFT algorithm executed by a physical computing system with physical limitations such as the need to round off some numbers. More noise. In fact, it's a tribute to modern technology that these noise components stay low-level.

As you can see, a lot of things separate the realities of measurement and digital analysis from the nicities of pure theory. Most of these things are discussed in greater detail in the next chapter. For now, however, let's just accept the fact that what appears in the FFT results, although unexpected, is probably valid or at least explainable.

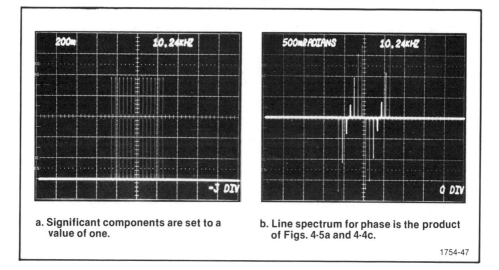

a. Significant components are set to a value of one.

b. Line spectrum for phase is the product of Figs. 4-5a and 4-4c.

1754-47

Figure 5-5 Interpreting phase from a periodic point of view requires looking only at the significant frequency components.

Nevertheless, the filtered square wave in Fig. 5–4a is certainly appealing to a continuous, periodic point of view. However, to take a periodic point of view for interpreting phase results, you must consider phase to be valid only at the frequencies of each spectral line in the magnitude display. With that in mind, the significant frequency components can be picked from the magnitude spectrum. Then they can be placed in an array and given a value of 1, while all other array elements are set to zero. The resulting array looks like Fig. 5–5a. Then, by element-by-element multiplication of the array of Fig. 5–5a with the phase array in Fig. 5–4c, a line spectrum for phase can be obtained (Fig. 5–5b). This line spectrum for phase and the magnitude spectrum in Fig. 5–4b are what you want to look at if you wish to interpret the frequency domain of the filtered square wave from a periodic point of view, a point of view relating to the Fourier series.

But, as has been pointed out in previous discussion, a nonperiodic point of view may be more appropriate for some analysis situations. Take Fig. 5–6a, for example. The pulse there certainly appeals to a nonperiodic point of view. You would probably want to interpret its frequency domain in terms of what would be expected from the Fourier integral.

The frequency domain of the band-limited pulse in Fig. 5–6a is shown in Fig. 5–6b and c. In particular, notice the phase display in Fig. 5–6c. It is well ordered, not like Fig. 5–4c, and certainly appears to agree with what is expected from the nonperiodic point of view. In fact, compare the frequency domain of Fig. 5–6 with the theoretical frequency domain of the Fourier integral in Fig. 3–7c. Notice how closely they match.

In fact, the only apparent difference from the integral results in Fig. 5–6 is in

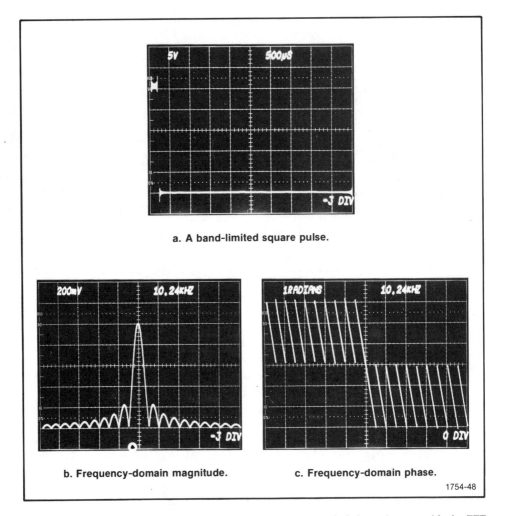

a. A band-limited square pulse.

b. Frequency-domain magnitude.   c. Frequency-domain phase.

1754-48

Figure 5–6 A case where the nonperiodic point of view is a natural choice and agrees with the FFT results.

the peak amplitude of the magnitude spectrum. Where the Fourier integral would place it at $VT_0 = 25 \times 200 \times 10^{-6} = 5 \times 10^{-3}$, the FFT places it at a value of 1. The reason for this rests in what is referred to as *the assumed periodicity of the FFT*. With assumed periodicity, it's as if the square pulse is actually a pulse train with a period equal to the window length. Assumed periodicity is discussed in detail later, but for now it causes the magnitude to be scaled to the mean value of the pulse over the window length. This mean value is the a_0 term of the Fourier series and can be computed for Fig. 5–6a as $VT_0/T = 5 \times 10^{-3}/(10 \times 500 \times 10^{-6}) = 1$, where T is the length of the acquisition window (500 microseconds per division for 10 divisions). So the results in Fig. 5–6 are completely predictable after all.

The example of Fig. 5–6 certainly supports the nonperiodic point of view for interpreting DFT and FFT results. And this should be expected, since the DFT is derivable from the Fourier integral. But remember that the Fourier integral is also derivable from the Fourier series. So a periodic point of view is also applicable to the DFT and FFT as long as you take into account the signal modifications occurring through the chain of derivation from the series, through the integral, to the DFT.

Continuous Phase.   Up to this point, phase has only been illustrated in terms of modulo 2π phase; that is, phase has been restricted to a range of $\pm\pi$, such as shown in Fig. 5–6c. This is sufficient for most interpretations. But phase can in theory and practice exceed the $\pm\pi$ range. This is shown in Fig. 5–7 (p. 88).

In terms of analyzing waveforms made up of many sinusoids, the phase of each frequency component is determined by the choice of zero time. In Fig. 5–8a (p. 89), for example, a pulse is shown arranged with time zero so that the pulse is an even function of time. The pulse is symmetric about time zero; therefore, it is made up solely of cosine terms. A heavy dot marks a reference point fixed to the pulse. The same reference is marked on two sinusoidal components in Fig. 5–8a that have been arbitrarily pulled from the pulse.

Now, as the pulse is shifted in time (Fig. 5–8b), each sinusoidal component undergoes the same time shift. The amplitude and time relationship of each component to the others is unchanged, since they must all still add up in the same manner to form the pulse. However, in terms of phase relation to time zero, each component must vary individually. This is because each component is of a different period, but each is time shifted by the same amount, and phase is the ratio of the time shift to the component's period (phase $= -360° \times$ shift/period). As indicated by the two arbitrary components in Fig. 5–8b, a higher frequency component, because of its shorter period, gains phase at a greater rate than one of lower frequency.

The pulse previously shown in Fig. 5–6 provides a good example for illustrating the difference between modulo 2π phase displays and continuous phase displays. This pulse is shown again in Fig. 5–9a (p. 90) with its position shifted slightly in time. Since time zero is at the left edge of the display, the additional shift in Fig. 5–9a should be reflected in the frequency domain as an increase in each component's phase.

Figure 5–9b shows the frequency-domain magnitude and modulo 2π phase for the pulse in Fig. 5–9a. This phase display is certainly different from that in Fig. 5–6, but notice that it's still confined to a range of $\pm\pi$ radians. Fig. 5–9c shows the same frequency domain as Fig. 5–9b; however, notice that the phase now exceeds the $\pm\pi$ range. Here, phase is shown in a continuous format, not interrupted nor repeated every $\pm\pi$. Even more interesting is the fact that the phase in Fig. 5–6, even though it was computed as modulo 2π, would look no different computed as continuous phase. That's because of the pulse's location relative to time zero in Fig. 5–6—the time shift, or delay, there is not great enough to cause phase to exceed $\pm\pi$. However, with the increased time shift in Fig. 5–9a, phase does exceed the $\pm\pi$ range.

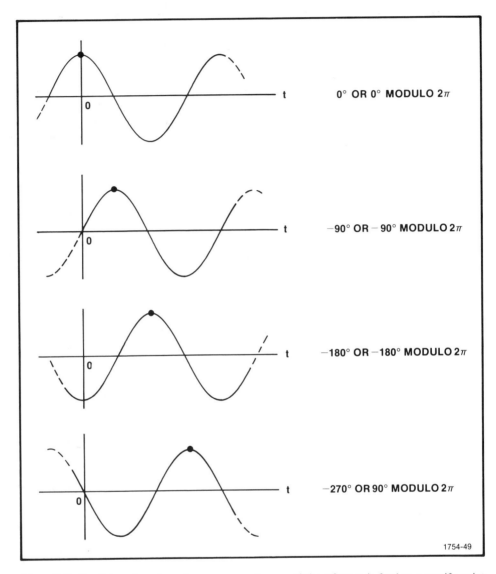

Figure 5-7 Specifying phase depends upon your reference. If the reference is fixed to a specific point on the shifted waveform, the phase related to time zero can exceed the $\pm\pi$ range.

Some FFT implementations provide polar results in modulo 2π phase and others provide continuous phase. Still others provide options for selecting polar results with either modulo 2π or continuous phase. It is important to keep in mind what kind of phase is being provided by the implementation you are using, so you can view the results accordingly.

It's also important to realize that modulo 2π phase and continuous phase are

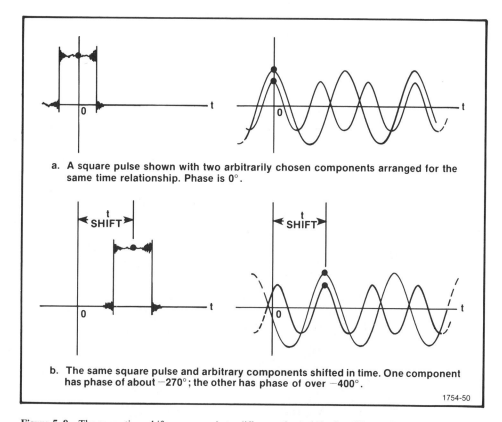

a. A square pulse shown with two arbitrarily chosen components arranged for the same time relationship. Phase is 0°.

b. The same square pulse and arbitrary components shifted in time. One component has phase of about −270°; the other has phase of over −400°.

1754-50

Figure 5–8 The same time shift corresponds to different phase shifts for different frequencies.

nothing more than two means of presenting the same information. Continuous phase presents the effects of time shifting (advance or delay) directly—360° of phase is represented as 360° of phase. On the other hand, modulo 2π phase views time shifting in the sense that a sinusoid with 0° phase (a cosine wave) appears no different from one shifted by 360°. Thus, phases of 360°, 720°, . . . are all represented as 0° in modulo 2π phase. This same idea applies to all phases in excess of $\pm\pi$ radians. They can be represented in the $\pm\pi$ range without changing the appearance of the sinusoid around time zero.

As a final note, there is another interesting thing that should be pointed out in Figs. 5–6 and 5–9. This isn't something new, but rather something in the manner of reviewing and reinforcing some theory covered in Chapter 3. Look at the pulses in Figs. 5–6a and 5–9a. The only difference between them is their location in time. Now look at the frequency-domain magnitudes and phases in Figs. 4–6b and c and 5–9b. Notice that there's absolutely no difference in the two magnitude displays. The phase displays, however, are markedly different! This points out again the Fourier transform property that time shifting only affects phase angle.

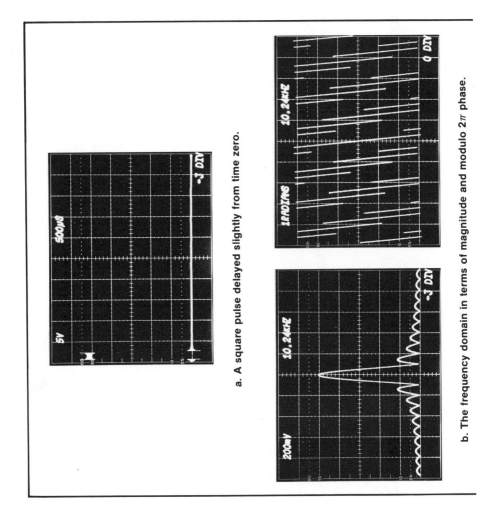

a. A square pulse delayed slightly from time zero.

b. The frequency domain in terms of magnitude and modulo 2π phase.

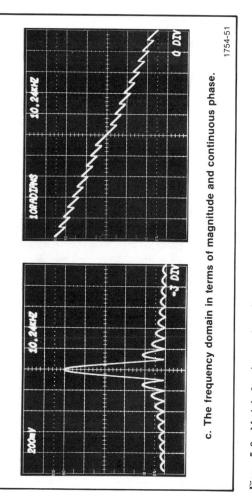

c. The frequency domain in terms of magnitude and continuous phase.

1754-51

Figure 5-9 Modulo 2π phase and continuous phase are two different ways of looking at the same thing.

Chapter 6

Understanding FFT Results— Recognizing the Realities of Digital Fourier Analysis

When an analysis technique is based heavily on theory, it is natural to let theory bias interpretations of the results. This is exactly as it should be, as long as theory is completely followed and precisely applied. Too often, though, results are interpreted in light of only the more general concepts. A rush to fill in the big picture leaves the details behind, and only a rough sketch is the result. And, too, there is often a tendency to categorize things—to place them under neat labels allowing easy prediction of results. For example, a 10-ohm resistor is only a 10-ohm resistor for convenience. It fits neatly into Ohm's law, $E = IR$, but in reality I may be a high-frequency current and R, a wire-bound resistor. Then the voltage drop, E, must be interpreted in light of the whole theory of conduction—capacitances, inductances, thermal effects—all must be considered.

Failing to be thorough and precise in using theory to predict or interpret results causes analysis errors. Sometimes the errors are in the results, but more often the errors are simply misinterpretations of the results.

In the case of the FFT algorithm, theory is applied thoroughly and precisely to the analysis of digitized waveforms. For the data supplied, the results are what they should be. And when the results aren't what we think they should be, it is most often a case of incomplete interpretation. Some feature of the waveform, maybe even a subtle one, has been overlooked. Or possibly some of the properties of windowing and digitizing have been overlooked or their implications not fully understood. Whatever the case, most FFT errors are easily explained or corrected by careful attention to detail.

There are two major classes of details that you should be aware of when you interpret FFT results. The first class of details concerns the waveform itself. What is being transformed? The second class of details concerns what happens to the waveform in preparing it for transformation by the FFT algorithm. What are the effects of changing an analog waveform to digital data?

Let's take a look at these two classes of details and explore their significance in terms of the FFT.

WHAT IS BEING TRANSFORMED?

Square Waves May Not Be Square Waves. When we look at a real-life waveform, we can often predict at least some of its frequency-domain features before actually transforming it to the frequency domain. For example, if it's a repetitive waveform, its period fixes the fundamental frequency. And if it's nonsinusoidal, we know that frequencies other than the fundamental are present as multiples of the fundamental.

Of course, if the waveform looks like a standard waveform, much more can be predicted. For example, if it looks like a square wave, we can go to Table 2–1 and see what frequency-domain components the Fourier series defines. From this, we should expect to see the fundamental and the odd harmonics when the waveform is transformed. Also, because we know the Fourier series for a square wave, we can say something about the expected amplitudes of these harmonics.

However, when the waveform is transformed, we shouldn't be surprised if the results don't exactly match our predictions. In the case of a waveform that looks like a square wave, predictions based on the Fourier series for an ideal square wave should not be taken too seriously. There simply are no ideal square waves in real life. Instantaneous rise and fall times, perfect symmetry, absolute steady amplitudes, all of the things that make an ideal square wave, aren't fully attainable in real-life waveforms. The distortions of real life, no matter how small, do affect the frequency domain.

Look at the waveform in Fig. 6–1a, for example. It looks like a square wave. And we could make some predictions based on the Fourier series for an ideal square wave. But look at the waveform again. How close does it really come to being an ideal square wave?

If you look closely enough, you can see that the waveform in Fig. 6–1a is really not an ideal square wave. This is confirmed by looking at its frequency-domain magnitude as obtained by the FFT (Fig. 6–1b). There the fundamental frequency (closest to the center and greatest in amplitude) has a frequency of about 12 kHz. With care, the third harmonic can be picked out at 36 kHz, the fifth harmonic at 60 kHz, and so on. But in doing this, notice that there are significant components between these odd harmonics. In fact, these extra components are even harmonics.

If the Waveform in Fig. 6–1a were an ideal square wave, these even harmonics wouldn't exist. However, they do exist because the waveform in Fig. 6–1a is not symmetric. The duration of the positive peak is slightly longer than the duration of the negative peak.

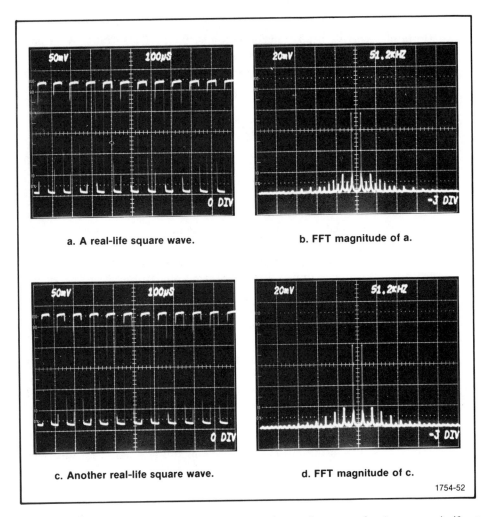

a. A real-life square wave.   b. FFT magnitude of a.

c. Another real-life square wave.   d. FFT magnitude of c.

1754-52

Figure 6–1 Barely perceptible differences in time-domain waveforms can often be seen as significant differences in the frequency domain.

Now look at the waveform in Fig. 6–1c. Except for a slight change in amplitude, it really doesn't look too different from the waveform in Fig. 6–1a. But look at the difference in the frequency-domain magnitude (Fig. 6–1d). The odd harmonics are now much clearer; the even harmonics having dropped significantly. The even harmonics are still visible, however, and their presence indicates that the waveform in Fig. 6–1c is still not quite symmetric.

The whole thrust of Fig. 6–1 is that real-life waveforms may look close to ideal, but they cannot always be considered ideal. For the purpose of general description, it is convenient to call the waveforms in Fig. 6–1 *square waves*. For the purposes of Fourier analysis, however, you must be very precise in how you define a waveform. If you are not precise or if you predict results on the basis of the ideal instead of the real, you might be surprised by the FFT results.

If you are surprised by the frequency-domain results of the FFT, go back and look closely at the original time-domain waveform. Ask yourself, What, exactly, is being transformed?

Time-Domain Noise Transforms to Frequency-Domain Noise.

Noise is a constantly present physical phenomenon. It has a multitude of sources that combine to produce an ambient condition of interference. And, in the case of generating and acquiring waveforms, the effects of additive noise cannot be discounted.

Quite often, proper shielding can reduce the effects of noise. But there are practical limits to shielding. There is always going to be some noise leakage, and where low-level signals are concerned, this noise leakage may contribute significantly to signal degradation. Also, there are those cases where noise is added to the signal before shielding can be applied. Consider radar and sonar transmissions. The transmitters and receivers can be shielded, but little can be done to prevent noise addition over the open-air transmission path.

When a noisy signal is acquired for Fourier analysis—for example, Fig. 6–2a—you must consider the noise as part of the signal. Often, noise is additive, and the acquired signal is the sum of the noise-free signal and the noise. In terms of the FFT, you should expect to see the transform of the noise-free signal summed with the transform of the noise (Fig. 6–2b). This is in keeping with the linearity property of the Fourier transform.

What constitutes an intolerable level of noise depends upon what you're looking for. If you're only looking for the general shape of things, then quite a bit of noise can be tolerated. In Fig. 6–2a and b, for example, the general shapes of the time-domain waveform and its frequency-domain magnitude are discernible through the noise. On the other hand, this same level of noise is intolerable when more precise information is desired. Then something similar to the waveforms in Fig. 6–2c and d is needed.

Noise is still present in Fig. 6–2c and d, but its level with respect to the signal is certainly much more tolerable. In the case of transient events, such improvements in noise level must be achieved through shielding or some kind of filtering. However, when the signal is repetitive and the noise random, signal averaging can substantially improve the signal-to-noise ratio. In the case of Fig. 6–2, the time-domain waveform is a single pulse from a train of pulses, and the improvement shown in Fig. 6–2c is obtained by signal averaging. When the averaged waveform is transformed to the frequency domain, the same type of improvement is seen there (see Fig. 6–2d).

The technique of signal averaging is straightforward. A repetitive signal is acquired a number of times, and each acquisition is added to the last. Then the sum is divided by the number of acquisitions. The result is an average of the acquired signals in the manner indicated in Fig. 6–3.

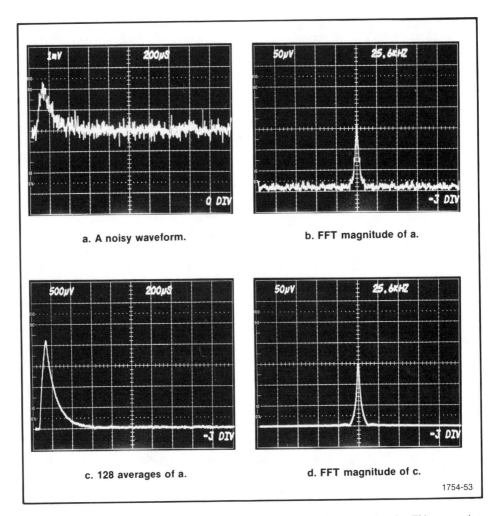

a. A noisy waveform.

b. FFT magnitude of a.

c. 128 averages of a.

d. FFT magnitude of c.

1754-53

Figure 6-2 Noise adds uncertainty to both the time domain and frequency domain. This uncertainty can be reduced by proper shielding and, in the case of repetitive signals, signal averaging.

Now, since ambient noise tends to be random and to have a mean of zero, the contribution of noise to the average is reduced. And, since the signal of interest is repetitive, averaging strengthens its contribution. As more acquisitions are averaged, the noise contribution is further reduced and the signal further reinforced. For truly mean-zero noise, the improvement in signal-to-noise ratio for M averages is $\sqrt{M}$. When M is expressed as a power of two, the improvement corresponds to 3 dB per power of two. For example, $2^7 = 128$ averages corresponds to a 7×3 dB $= 21$ dB improvement in signal-to-noise ratio. If more improvement is needed, the number of averages, M, is increased.

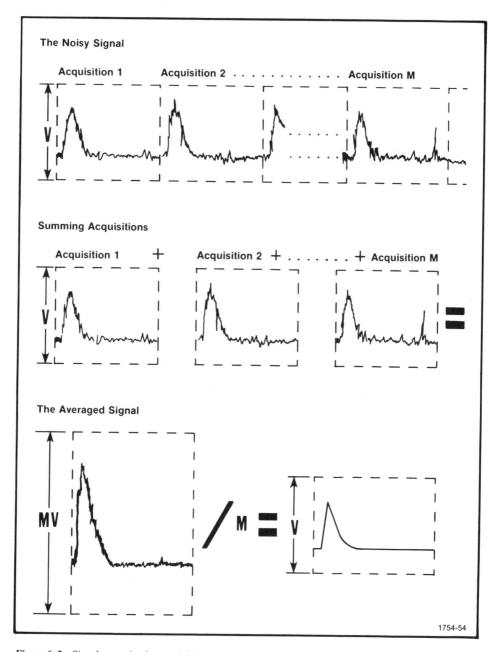

Figure 6–3 Signal averaging is a straightforward technique for pulling repetitive waveforms out of random, mean-zero noise.

WHAT ARE THE EFFECTS
OF ANALOG-TO-DIGITAL CONVERSION?

When an analog waveform is to be transformed to the frequency domain by the FFT, it must first be converted to a digital representation. The steps in this conversion include acquiring the waveform through analog circuitry, sampling it through a data window, and finally converting the samples to digital words representing the waveform. If the FFT results are interpreted strictly from the viewpoint of the digitized samples, the results are exact. However, it is more likely that you'll want to interpret the results from the viewpoint of the analog waveform. Then the changes that the waveform undergoes in analog-to-digital conversion must be considered in the interpretation.

Conversion Noise. For the most part, noise generated by the analog-to-digital converter can virtually be eliminated through proper design procedures. There are two major noise sources, however, that are more closely process related than hardware related. These are time jitter during acquisition and quantizing noise during digitizing.

The first of these two, *time jitter*, occurs, for example, when the analog waveform activates a level trigger that gates the acquisition window on. For various reasons, noise on the signal being one, subsequent windows of the same waveform may be triggered at slightly different points on the waveform. This results in the waveform losing horizontal (time) stability in the window. This jittering back and forth is sometimes seen on oscilloscope displays (Fig. 6–4a) when the oscilloscope trigger level is set at an unstable point on the waveform or when waveform variations (noise, for example) cause the trigger point to shift on the waveform.

For single-shot events or sampling techniques where the waveform is digitized in a single window, trigger jitter isn't generally a problem. The waveform is triggered once and acquired rather than acquired over several triggered sweeps. The result is that any slight time shift in the window (or trigger) simply translates to the same time shift in each sample location on the waveform. However, if individual sample windows jitter relative to one another (as might be caused by sample clock instability), a problem exists for single-shot acquisitions. Such sample jittering is best handled through careful digitizer design and specification.

Time jitter from triggering instabilities is a greater threat when repetitive waveforms are being acquired and digitized from samples taken over a number of windows. (Several sweeps of the waveform are required to build up a full complement of samples.) The effect is a shifting of some samples on the waveform with respect to other samples. This causes the stored waveform to take on the appearance shown in Fig. 6–4b.

Notice in Fig. 6–4b how the sampling error from jitter resembles the random noise shown in Fig. 6–2a. And, if the waveform in Fig. 6–4b is transformed to the frequency domain, the effect is the same as that caused by additive noise. Although the amplitudes of the variations due to jitter are related to signal slope (dv/dt), the variations themselves are often random in occurrence, like additive noise, and are apt to have a zero mean over all. Because of these characteristics, signal averaging

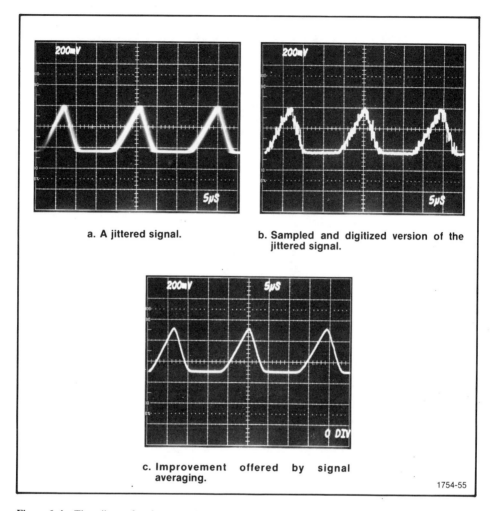

a. A jittered signal.

b. Sampled and digitized version of the jittered signal.

c. Improvement offered by signal averaging.

1754-55

Figure 6–4 Time jitter often has a random, noise-like effect that can be reduced by signal averaging.

can be used to reduce the effects of time jitter. This is shown in Fig. 6–4b and c, where 512 averages were used to gain the signal-to-noise improvement indicated by Fig. 6–4c.

There is another type of conversion noise that you should know about too. This is *quantizing noise*, also referred to as *quantizing error*. In general, quantizing noise is a nonrandom noise of fixed level. Therefore, it cannot, in theory, be reduced by signal averaging. In practice, however, the effects of quantizing noise relative to the acquired signal can be reduced in some cases.

To understand the source of quantizing noise, consider digitizing a ramp. This is shown on a limited scale in Fig. 6–5, where the dots represent possible values and sample locations provided by a digitizer. Figure 6–5 is referred to as being on

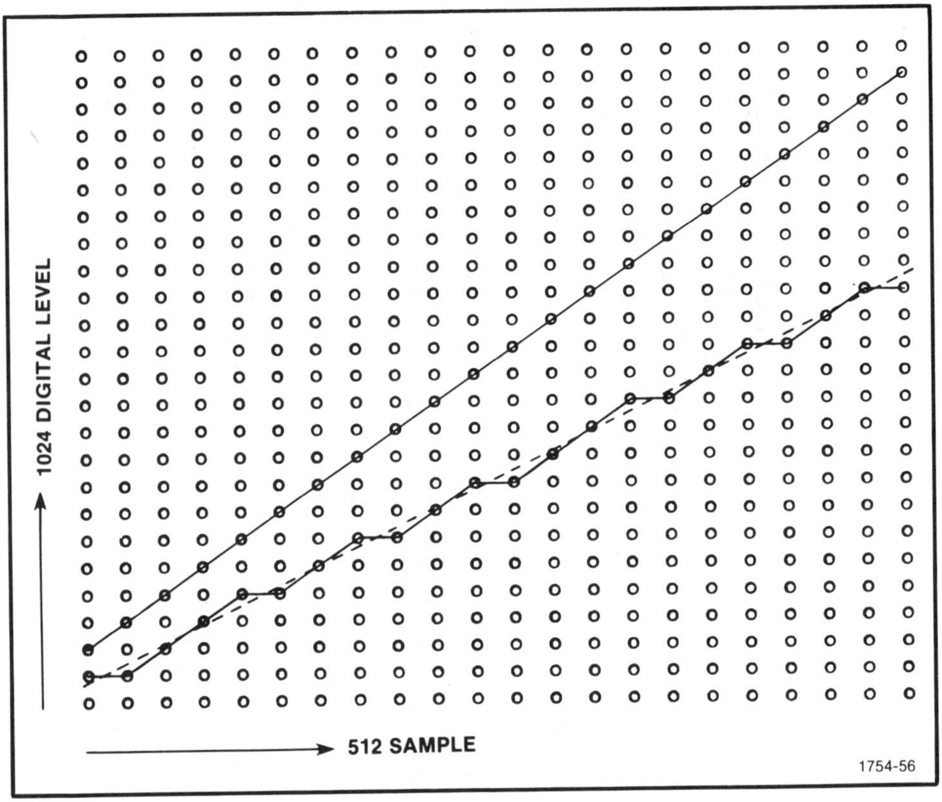

Figure 6–5 Quantizing error is inherent in digitizing. The top ramp lies on the digital levels. The bottom ramp (dotted) lies between digital levels, and its digital description (solid line) suffers quantizing error.

a limited scale because an actual digitizer generally provides more possible values and locations than shown there. For example, a typical waveform digitizer might sample at 512 locations horizontally in the time window and provide conversion of vertical values to 10 bits. Ten bits means that there are $2^{10} = 1024$ digital levels for expressing the amplitude at each sample location. The combination of sample points and possible amplitude levels can be considered to compose a gridwork, the points of which represent the allowable values for describing a waveform. A portion of such a grid is shown in Fig. 6–5.

There are two ramps shown on the grid in Fig. 6–5. Notice that the top ramp has values falling exactly on the available digital levels. When this ramp is sampled and digitized, the digital representation exactly matches the ramp. Now look at the bottom ramp in Fig. 6–5 (dotted line). In a number of places, its samples fall between the allowable digital levels, so the digitizer must make a decision as to which level it assigns to the sample. As a result of this electronic rounding off, the digital representation of the lower ramp in Fig. 6–5 follows the path shown by the solid line.

The slight deviations from the dotted line in Fig. 6–5 are quantizing error and are inherent in any digitizer of finite word length. The amount of this error at any sample point is no more than half of a digital level positively or negatively and may under certain conditions occur randomly over the block of waveform samples. The expression "may under certain conditions occur randomly" is used with caution here because theory classifies quantizing error as nonrandom. In other words, given exactly the same signal acquired exactly the same way every time with a rock-solid digitizer, quantizing error occurs to the same degree at the same place on each acquisition of the signal.

In practice, however, how many things are rock solid and the same every time? Jitter the ramp slightly, and the quantizing error jitters about relative to the ramp. If the jitter is random, then the quantizing error relative to the ramp appears to be random, just like noise. Thus, quantizing error is often referred to not as error but as *quantizing noise.*

Since quantizing noise by itself doesn't exceed half a digital level in peak value (one digital level peak to peak), it is generally low-level compared with other types of noise. However, it can become high-level noise compared with the signal. To understand this, consider acquiring a waveform so it ranges over only 100 of the available 1024 digital levels of a 10-bit converter. The quantizing error is then half a digital level out of the 100 levels used. Now consider acquiring the waveform through an input amplifier so it ranges over all 1024 digital levels. A reduction in the relative effect of quantizing noise of more than 10 times is achieved in this case by simply exercising the full range of the analog-to-digital converter.

Even if the full range of an analog-to-digital converter is used, there's still a possibility of half a vertical level of quantizing noise. Increasing digitizer word length is one way to further decrease the amplitude of this noise relative to the signal. For example, a 16-bit digitizer is more finely grained than a 10-bit digitizer. But there are practical limits to digitizer word length.

Signal averaging is another approach that can sometimes reduce quantizing noise. Any success, however, depends upon two things. First, there must be enough other additional noise—either additive noise or jitter—to inject some randomness into the quantizing error and to raise the total noise level above two or three digital levels. The key thing is that the quantizing error gets jittered randomly relative to the signal. Second, the signal averaging must be done with a computing device having more resolution (less quantizing error) than the digitizer being used. For example, using a 16-bit minicomputer to signal average the data acquired by a 10-bit waveform digitizer can result, in some cases, in the averaged waveform having a better signal-to-noise ratio than would be expected from the 10-bit digitizer.

Ultimately, however, noise can never be fully removed from a digitized signal. You should expect it to be present at some level in all cases of digital signal processing. This noise may have become part of the signal before it was acquired (ambient, additive noise), or it may become part of the signal during analog-to-digital conversion (jitter, quantizing, and other sources). In most cases, the noise level can be reduced significantly by shielding or signal averaging. But it can never be removed completely. Therefore, noise is always present to some degree in the FFT results too. With care, its level with respect to actual signal components can be kept small enough to be considered zero. However, when further processing is done—computing phase from the real and imaginary parts, for example—don't forget that low-level noise components can make significant contributions to the results. A small value divided by a small value is the same as a big value divided by a big value.

Windowing Can Cause Leakage. Let's begin talking about leakage by considering a pure cosine wave and its FFT magnitude. We'll consider a cosine wave because it has zero phase (a zero imaginary part); thus we can concentrate on just the FFT magnitude for what needs to be said about leakage.

The pure cosine wave is shown in Fig. 6–6a. Its array elements were generated by a computer program rather than by acquiring and digitizing a waveform from a signal generator. This mathematically generated waveform is preferred for this demonstration because each waveform element can be precisely controlled, at least within the computational limits of computer word length. In the case of Fig. 6–6a, the cosine wave has been generated so that exactly 10 cycles appear over the 10-sec window length. Also, the cosine wave has an exact amplitude of 1 V.

The FFT magnitude of the cosine wave is shown in Fig. 6–6b. Its major parameters, taken from the array of computed Fourier coefficients, are printed out on the display. Notice that they describe exactly what we would expect for the cosine wave in Fig. 6–6a. The FFT magnitude indicates a cosine wave existing at exactly 1 Hz with an amplitude of 1 V. (Remember, half the energy is in the positive frequency domain and half in the negative frequency domain.)

Now, let's change the frequency of the cosine wave just a little. Let's increase it so that exactly 10.5 cycles appear in the 10-sec window (1.05 Hz). This new cosine wave is shown in Fig. 6–6c and has an amplitude of exactly 1 V. Its FFT magnitude is shown in Fig. 6–6d. Again, the major parameters of the FFT magnitude are printed out at the bottom of the display. They indicate that the cosine wave has an amplitude of 0.6508 V and a frequency of 1.1 Hz.

What happened to our 1-V, 1.05-Hz cosine wave? Leakage is what happened. Compare Fig. 6–6b and d. Figure 6–6b is the line spectrum expected for a cosine wave, whereas Fig. 6–6d is not. Figure 6–6d is more like something you'd expect from transforming the cosine wave over a short interval with the Fourier integral. In fact, that is what has been done in both cases, the effect in one case being to include 10 cycles and in the other case 10.5 cycles. Also, remember that limiting the integral to a short interval or window is the same as multiplying the signal by a square pulse having a width equal to the interval of Fourier integration. The effect of the $(\sin x)/x$ frequency-domain function for such a rectangular data window is very apparent in Fig. 6–6d. There the data window has caused noticeable widening at the bases of the two spectral components. This widening of the primary components can be thought of as leakage of primary power into adjacent frequencies.

But why should Fig. 6–6d be any different from Fig. 6–6b? After all, both magnitude spectra are for cosine waves of the same amplitude and very nearly the same frequency. Why should the discrete transform treat them differently? In answer,

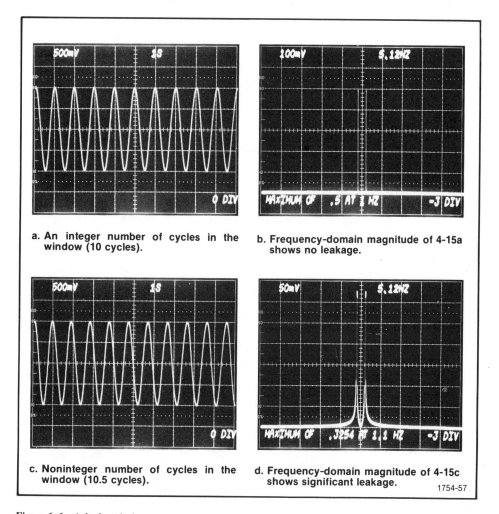

a. An integer number of cycles in the window (10 cycles).

b. Frequency-domain magnitude of 4-15a shows no leakage.

c. Noninteger number of cycles in the window (10.5 cycles).

d. Frequency-domain magnitude of 4-15c shows significant leakage.

1754-57

Figure 6–6   A look at leakage.

an integer number of cycles is acquired within the window in one case; and in the other case, a noninteger number of cycles is acquired.

To understand the effect on leakage of the number of cycles in the window, let's look closely at what happens when each cosine wave is prepared for transformation by the FFT. This is best done through Figs. 6–7 and 6–8. In both of these figures, cosine waves and sampling rates of lower frequencies are used for illustrative convenience. Positional relationships to the window are maintained, however, so the same concepts apply.

Beginning in Fig. 6–7a, a cosine wave is shown in both the time domain and the frequency domain. This is the ideal. The cosine wave is exact, and its amplitude

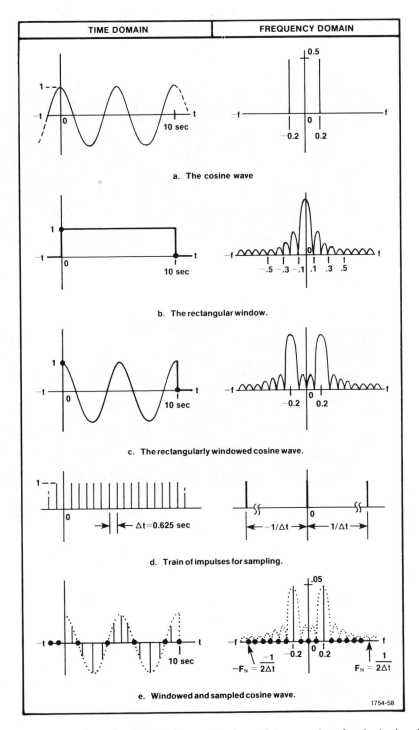

Figure 6–7 Process of transforming a cosine wave having an integer number of cycles in the window.

and frequency are reflected exactly in the frequency-domain line spectrum. (Since phase is zero, a phase spectrum isn't indicated.)

The rectangular window corresponding to the time-domain displays areas in Figs. 6–6a and c is shown in Fig. 6–7b. It is defined in the time domain such that it is zero for all negative time and goes to a value of unity at time zero. It then remains at unity amplitude out to 10 sec. There it falls instantaneously so its defined amplitude at 10 sec is zero. In the frequency domain, the magnitude of the rectangular window is the familiar $(\sin x)/x$ of a square pulse. Because of the time width of the window, the magnitude diagram shows a zero at every 0.1 Hz (reciprocal of window width). The phase for this particular window is nonzero; however, its combination with other zero-phase waveforms in this example makes phase consideration unnecessary for now.

The first step of analog-to-digital conversion, whether through hardware acquisition or mathematical waveform generation, is shown in Fig. 6–7c. This is a limiting of our view of the waveform to a finite time interval, a window. In the time domain, this is mathematically equivalent to multiplying the rectangular window in Fig. 6–7b with the cosine wave in Fig. 6–7a. In the frequency domain, this time-domain multiplication corresponds to convolving the frequency-domain magnitude in Fig. 6–7a with that from 6–7b. (Since the cosine wave has zero phase, its frequency-domain convolution with the window results in a zero phase function.) Notice that the maximum of the two $(\sin x)/x$ major lobes in Fig. 6–7c falls directly on the 0.2-Hz frequency of the cosine wave. Because of the pulse nature of the window, however, the frequency domain of the windowed cosine wave has now become a continuous spectrum instead of a line spectrum.

The windowed cosine wave in Fig. 6–7c still hasn't been converted to digital form yet. It's still an unsampled analog waveform.

To finish the conversion, we must limit our view of the waveform to specific points within the window's unity amplitude duration. This further limiting to specific points is referred to as *sampling* and can be represented as a train of impulses similar to that shown in Fig. 6–7d. The impulses in this train define the discrete time-domain points to be considered in the transformation.

For ease of illustration, Fig. 6–7d shows only 16 samples occurring within the time span of the rectangular window. Sample 17 falls on the zeroing edge of the window. Each sample is separated by a Δt of 0.625 seconds. This is substantially fewer samples than used in Fig. 6–6. The window there included 512 samples at a spacing of $\Delta t = 0.01953$ sec; however, the example of Fig. 6–7 doesn't require such resolution.

Figure 6–7d also shows the frequency-domain magnitude corresponding to the impulse train (phase is zero). Notice that there is an impulse at 0 Hz and that impulses occur every $\pm 1/\Delta t$ Hz from there. For the current discussion, the only impulse of concern is the one occurring at 0 Hz. The other frequency-domain impulses play a significant part in later discussion of what is often referred to as the *assumed periodicity of the DFT and FFT*.

Time-domain sampling of the windowed waveform corresponds mathematically to multiplying the windowed waveform in Fig. 6–7c by the impulse train in Fig. 6–7d. In the frequency domain, this corresponds to convolution of the windowed waveform's frequency domain with the frequency domain of the impulse train. The results of this sampling are shown in Fig. 6–7e, where the waveform envelopes are suggested by the dotted lines.

The time-domain samples in Fig. 6–7e are the waveform amplitudes that are expressed digitally and transformed to the frequency domain by the FFT. There are 16 samples shown in the time domain of Fig. 6–7e. The first valid sample is considered to be at time zero, since that is where the window becomes defined as having a value of one. The last valid sample is located 15 samples from this. The next sample, sample 17, is not valid since it occurs at the right edge of window, which is zero by definition.

Transforming these 16 time-domain samples to the frequency domain results in 16 complex samples of the frequency domain. The arrangement of these samples on the envelope of the frequency-domain magnitude is shown in Fig. 6–7e. They span a frequency range of $1/\Delta t = 1.6$ Hz centered on 0 Hz. Seven samples are shown in the positive frequency domain, 1 is shown at 0 Hz, and 8 are shown in the negative frequency domain. The last sample shown in the negative frequency domain (left edge) is referred to as the *Nyquist frequency* (F_N) and is equal to $1/2\Delta t$. The Nyquist frequency is the highest frequency that can be defined by the $1/\Delta t$ sampling rate. (Note that, in other display formats, the Nyquist frequency may be shown at the positive frequency extreme. This certainly is the case when only the positive frequencies are displayed.)

From the standpoint of leakage error, the location of the frequency-domain samples relative to the envelope of the frequency-domain magnitude is most important. Notice in Fig. 6–7e that the maxima of the two envelope peaks correspond exactly to the sample points at ± 0.2 Hz. This is the exact frequency of the cosine wave in Fig. 6–7a. Also, notice that the remaining samples fall on the zeros of the magnitude envelope. The result is, in essence, the exact line spectrum for the cosine wave.

Leakage doesn't appear in Fig. 6–7e because an integer number of cosine wave cycles is enclosed in the sampling window. This means that the cosine wave in Fig. 6–7 is harmonically related to the window length. With the sample arrangement shown, the cosine wave's frequency falls exactly on a sample point. All other sample points occur at the zeros of the magnitude envelope. If we illustrated the same things with the 512 samples used in Fig. 6–6, the same effect would be seen. The only difference would be that the frequency range covered by the transform would be 51.2 Hz. The spacing of the frequency samples, however, would still be at 0.1 Hz in either direction from 0 Hz.

Now we know why leakage isn't seen in Fig. 6–6b. It's simply because the repetitive waveform is harmonically related to the window length; that is, there's an integer number of cycles in the window.

But what about Fig. 6–6d? It represents a noninteger number of cycles in the window. How does this bring leakage into the results? To answer this, think about what would happen if the frequency of the repetitive waveform was such that it fell between frequency-domain sample points. Can you imagine what would happen then? Figure 6–8 will help you see the situation.

Figure 6–8 shows the exact same steps as in Fig. 6–7. This time, though, the windowing, sampling, and transformation process are carried out on a cosine wave having a frequency of 0.25 Hz instead of 0.2 Hz. This cosine wave and its frequency-domain line spectrum are shown in Fig. 6–8a. Following this, in Fig. 6–8b, are the time-domain window and its frequency-domain magnitude. Notice that the same 10-sec window from Fig. 6–7 is used. As was done in Fig. 6–7, the cosine wave and the window are multiplied. The product of the two is shown in Fig. 6–8c. Notice that the window contains a noninteger number of cosine cycles (2.5 cycles). The same step in Fig. 6–7 included an integer number of cycles (2 cycles).

Following the product of the window and cosine wave, the sampling train is shown in Fig. 6–8d. Again nothing has changed from Fig. 6–7. The sampling train has exactly the same position and spacing between sample points. And it is applied to the windowed waveform in the same manner as described for Fig. 6–7.

Figure 6–8e shows where the effect of changing the cosine wave's frequency really becomes obvious. In particular, notice the position of the magnitude envelope relative to the samples. Notice how the two maxima of the magnitude envelope fall between the 0.2 and 0.3 Hz samples. This causes two samples to fall on either side of each major lobe maximum. Also, notice that all of the remaining frequency-domain samples now fall on the peaks of the side lobes instead of on the zeros of the magnitude envelope. This is called *leakage error*. It is the taking of power from components existing in the continuous waveform and giving power to frequency components that don't exist in the continuous waveform. Because of this leakage error, the discrete transform of the cosine wave is no longer the expected line spectrum. However, it should be pointed out that the discrete transform is exactly what it should be for the time-domain samples provided in Fig. 6–8e.

The same holds true for Fig. 6–6d. There is leakage error there too, if you choose to interpret the frequency domain in terms of the line spectrum for a periodic waveform. But if you interpret the results from a digital viewpoint, the results are correct for the 512 time-domain samples provided in Fig. 6–6c.

To reiterate, the leakage error seen in Figs. 6–6c and 6–8e comes about because the periods of the cosine waves are not harmonically related to window length. In each case, there is a noninteger number of cycles in the window, and in each case, leakage can be removed by adjusting the window length to include an integer number of cycles.

Figures 6–7 and 6–8 point out some of the basic properties of leakage. But there are some other things that should be pointed out too. For one, leakage is not restricted to the frequency-domain magnitude. In dealing with waveforms in general, leakage error, when it occurs, affects the imaginary part and the real part of the frequency domain to the same degree. Also, when converting to the polar form, leakage is carried through the conversion to affect both magnitude and phase.

Another thing you should realize is that leakage has the same relative effect on the harmonics of a nonsinusoidal, repetitive waveform as it has on the fundamental

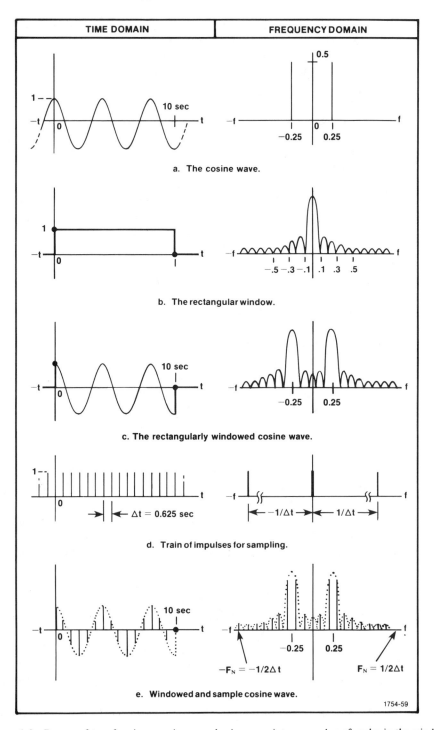

a. The cosine wave.

b. The rectangular window.

c. The rectangularly windowed cosine wave.

d. Train of impulses for sampling.

e. Windowed and sample cosine wave.

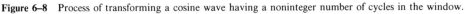

Figure 6–8 Process of transforming a cosine wave having a noninteger number of cycles in the window.

of that waveform. This follows from the fact that the degree of leakage is related to the harmonic relationship of the fundamental to the window length. In turn, the harmonics see the relationship to the window through their relation to the fundamental. For example, if the waveform is acquired so that an integer number of cycles is in the window, then each of its harmonics also has an integer number of cycles in the window. Each spectral component falls precisely on an array element in the frequency-domain arrays, rather than between elements.

With regard to nonrepetitive waveforms, specifically pulses, leakage is not a problem as long as the pulse is fully contained in the window. In other words, the pulse must rise from its base and return to its base within the window. If you acquire a pulse that is partially outside the window, then you have failed to fully define the pulse and should expect inconsistencies in the frequency domain.

Another type of nonrepetitive waveform must be dealt with too. This is the almost periodic waveform, which is made up of discrete frequencies that are not harmonically related. This type of waveform is often found in vibration studies where the waveform is the result of several vibration sources. Since one or more of its components may not be a harmonic of the others, all components cannot be harmonically related to the window length. Thus some components may exhibit leakage while others don't.

As a final note on leakage, the form that leakage takes depends on the form of the window. The leakage shown thus far has been of a form uniquely tied to a rectangular data window. By changing the shape of the window, as shown in Fig. 6–9, you can change the shape of the leakage. The properties of various window shapes and some guidelines for using them are given later in Chapter 7.

Assumed Periodicity—A Case of Even Becoming Odd. In order to demonstrate the periodicity inherent in sampling and windowing and assumed by the FFT, let's return to the cosine waves used to introduce the topic of leakage. In one case, exactly 10 cycles of the cosine wave fill the window; in the other case, the window is filled by 10.5 cycles.

Now let's use the FFT again to transform the 10 cycles of cosine wave to the frequency domain. This is shown in Fig. 6–10a (p. 112). Notice that this time, however, the frequency domain is shown in rectangular form instead of the polar form used in discussing leakage.

The frequency domain in Fig. 6–10a is what we should expect from theory. Since a cosine wave is an even function, its frequency-domain function is a real and even function. This is shown in Fig. 6–10a by the nonzero real part and the effectively zero imaginary part.

Now let's use the FFT to transform the 10.5 cycles of cosine wave. Since we're still dealing with a cosine wave—an even function in theory—something similar to Fig. 6–10a might be expected. But this isn't the case! In fact, the FFT results for 10.5 cycles (Fig. 6–10b) come out exactly the opposite. The real part is effectively zero, and the imaginary part is nonzero and an odd function. Instead of being the

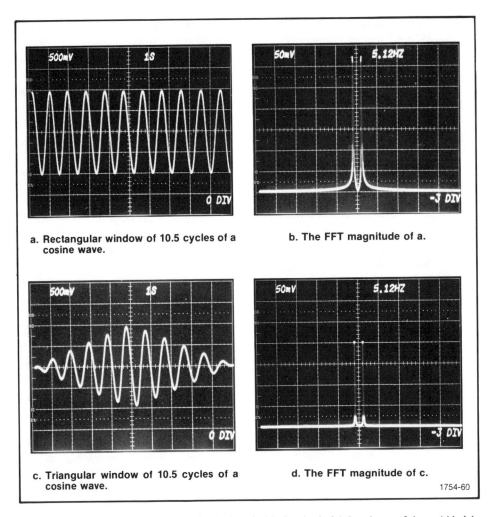

a. Rectangular window of 10.5 cycles of a cosine wave.

b. The FFT magnitude of a.

c. Triangular window of 10.5 cycles of a cosine wave.

d. The FFT magnitude of c.

1754-60

Figure 6–9 Window shape determines the degree of widening (main lobe) and rate of decay (side lobe size) of a spectral component.

frequency domain for an even function, Fig. 6–10b is the frequency domain for an odd function.

The paradox in Fig. 6–10 is easily explained if we think of the waveforms there as being continued beyond the windowed edges. But we can't totally ignore the window either. So imagine the waveforms being continued by repeating the window and its contents. In other words, duplicate the windowed cosine waves in Fig. 6–10. Then lay these duplicates end to end on either side of the original window, so that the information in each window repeats with a period equal to the window length. This is illustrated more clearly in Fig. 6–11 (p. 114). There, the original window is marked by solid lines and the repeated windows by dotted lines.

Notice in Fig. 6–11a how the windows containing exactly 10 cycles of the waveform allow the waveform to continue on in a cosinusoidal fashion. This is because the window edges fall at the same relative points on the waveform. Now notice how this doesn't happen for the 10.5 cycles of cosine wave (or any noninteger number of cycles). The window edges in Fig. 6–11b don't fall on the same relative points. Therefore, repeating the window for the case of 10.5 cycles results in an odd function of time. For the case of any other noninteger number of cycles (10.25 cycles, for example), repeating the window results in a function that is neither even nor odd but the sum of even and odd parts.

Figure 6–11 illustrates the assumed periodicity of the FFT. This periodicity isn't necessarily related to the windowed waveform, but it is related to the window. The window of data is assumed by the FFT to repeat periodically. Exactly how this comes about is illustrated in Fig. 6–12 (pp. 116–17).

Figure 6–12 begins at a point corresponding to Fig. 6–8d. That is, a cosine wave has been windowed so that 2.5 cycles are contained in the window. Figure 6–12a shows the time-domain impulse train representing sampling and the frequency-domain magnitude of this sampling train. The Δt-spaced impulses in the time domain and the corresponding $1/\Delta t$-spaced impulses in the frequency domain extend in both directions to infinity.

In Fig. 6–12b, the effects of time-domain sampling are shown. In the time domain, the windowed cosine wave of Fig. 6–8c is multiplied by the sampling train. This corresponds to convolution in the frequency domain. When the windowed cosine wave's frequency-domain magnitude is convolved with the $1/\Delta t$-spaced impulses in the frequency domain, the result is a repetition of the windowed cosine wave's frequency-domain magnitude at every $1/\Delta t$ Hz.

Since the FFT computes only discrete frequency-domain points corresponding to the time-domain samples, the FFT can be equated to frequency-domain sampling. Such a sampling train, with an impulse spacing of $\Delta f = 1/N\Delta t = 0.1$ Hz, is shown in the frequency domain by Fig. 6–12c. The time-domain function corresponding to this frequency-domain impulse train is also shown in Fig. 6–12c. It, too, is a train of impulses, but with a time spacing of $1/\Delta f = 10$ sec.

Figure 6–12d shows the result of frequency-domain sampling. The sampling corresponds to multiplying the frequency-domain sampling train of Fig. 6–12c with the frequency-domain magnitude of Fig. 6–12b. This multiplication corresponds to convolution in the time domain. Thus, the windowed and sampled cosine wave in Fig. 6–12b is convolved with the $1/\Delta f$ impulses in Fig. 6–12c. The result of this is repetition of the windowed and sampled cosine wave in the manner shown in Fig. 6–12d.

The actual data taken by windowing and sampling and the data computed by the FFT are enclosed in Fig. 6–12d by dashed lines. These data blocks are what actually fill the software arrays and what make up the instrument displays. In practice, however, more than 16 samples (as shown in the dotted box) are usually used for each array. The effective repetition of these data blocks comes about through sampling the original waveform. How the assumed periodicity of the FFT affects the periodicity

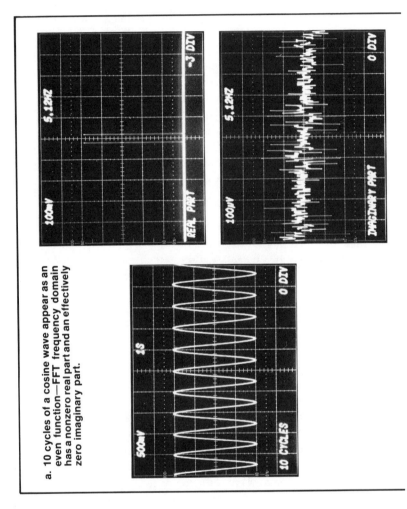

a. 10 cycles of a cosine wave appear as an even function—FFT frequency domain has a nonzero real part and an effectively zero imaginary part.

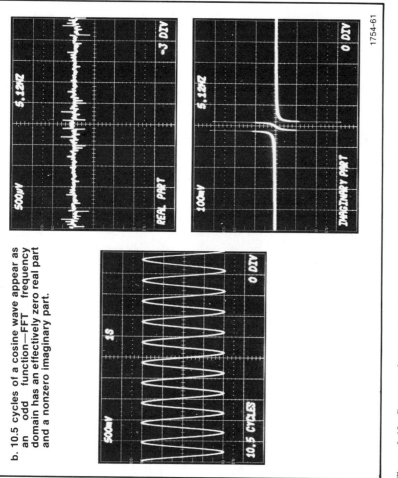

b. 10.5 cycles of a cosine wave appear as
   an odd function—FFT frequency
   domain has an effectively zero real part
   and a nonzero imaginary part.

1754-61

Figure 6–10 Because of assumed periodicity in the FFT, an even function can appear to be an odd function.

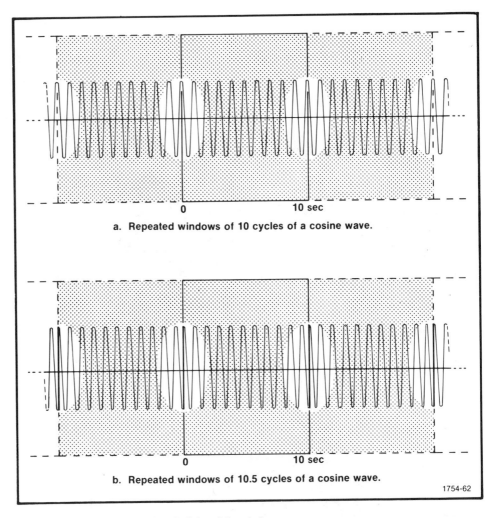

a. Repeated windows of 10 cycles of a cosine wave.

b. Repeated windows of 10.5 cycles of a cosine wave.

1754-62

Figure 6–11   The FFT assumes periodicity of the window.

of the original waveform depends upon how the samples are arranged on the original waveform. For the case where an integer number of cycles is acquired in the window, the assumed periodicity of the FFT is harmonically related to the waveform's period. If a noninteger number of cycles is acquired, the waveform's period isn't harmonically related to the assumed periodicity, and the situation of Fig. 6–10b occurs. In either case, the FFT results are correct, considering the effects of windowing and sampling. The error comes in when digital results are interpreted strictly from an analog point of view.

As a final note, assumed periodicity is independent of the waveform being acquired. Periodic repetition of the sampling window always occurs. Thus, an acquired and sampled nonperiodic waveform (a pulse) is also subject to assumed periodicity in the manner of Fig. 6–12. However, if the pulse is acquired so it is completely defined within the sampling window, the actual time-domain samples and the computed Fourier coefficients are a correct representation of the pulse. The assumed periodicity simply causes the information to appear duplicated beyond the window edges.

Aliasing—A Case of Incomplete Definition. Before any waveform can undergo digital signal processing, the waveform must be windowed and sampled. The rate at which the waveform is sampled determines how well it is defined and how close the discrete representation is to the analog original.

The rule governing proper sampling is referred to as the *Nyquist sampling theorem,* which states that *the sampling rate must be at least twice the frequency of the highest frequency component in the waveform being sampled.* In other words, there must be at least 2 samples per cycle for any frequency component you wish to define. If there are fewer—if the sampling rate is less than twice the highest frequency component—then aliasing occurs.

You can verify the Nyquist sampling theorem through a very simple experiment. It consists of sampling sinusoids of increasing frequency while maintaining a constant sampling rate. At some point, the sinusoid's frequency becomes such that samples occur at less than 2 per cycle. Then aliasing error can be seen. This is most easily observed in the frequency domain, where the spectral components can be observed to move out to the edges of the display as frequency is increased. When the Nyquist frequency is reached (the frequency where aliasing begins), the spectral components will have moved to the edges of the display. As the waveform frequency is further increased, the components fold around the edges of the FFT magnitude display and move back to lower frequency areas. This is aliasing. It is the representation of a high-frequency component by a lower-frequency component. The key points of this experiment are shown in Fig. 6–13 (pp. 118–19).

In Fig. 6–13a, 20 cycles of a cosine wave are included in the window. For 512 samples in the window, this gives about 25 samples per cycle, which is more than ample for defining the cosine wave. The FFT magnitude for this 25-samples-per-cycle condition is shown in Fig. 6–13a. It is what would be expected for the sampled cosine wave.

Now, in Fig. 6–13b, the frequency is increased so that 200 cycles appear in the window. This means there are a little over 2 samples per cycle. (In practice, it is best to use at least 3 samples per cycle on the highest frequency component to be certain of avoiding aliasing.) As can be seen by the time-domain representation, it is a little difficult to visualize the cosine wave when only 2.56 samples per cycle are used. Nevertheless, the samples do define the cosine wave. It is correctly indicated by the spectral components near the first and ninth divisions in the FFT magnitude display.

Another increase in frequency brings us, in Fig. 6–13c, to a condition of 2.016 samples per cycle. The spectral components in the FFT magnitude have moved nearly to the edges of the display. This is nearly the limit set by the Nyquist sampling theorem.

TIME DOMAIN

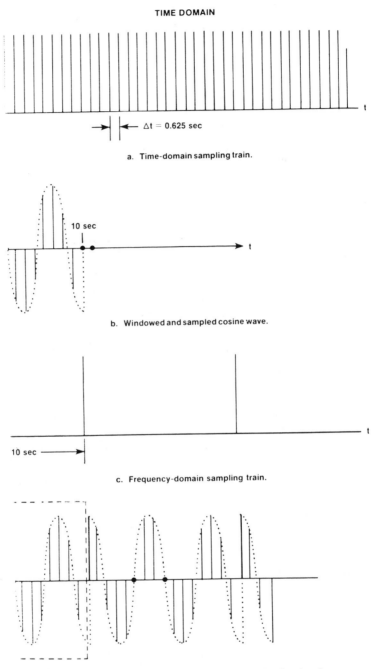

a. Time-domain sampling train.

b. Windowed and sampled cosine wave.

c. Frequency-domain sampling train.

d. The sampled frequency domain and corresponding time domain.

Figure 6–12 FFT periodicity is tied in with sampling.

FREQUENCY DOMAIN

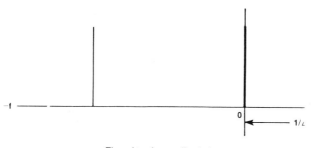

a. Time-domain sampling train.

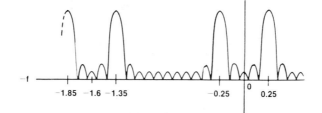

b. Windowed and sampled cosine wave.

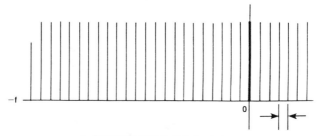

c. Frequency-domain sampling train.

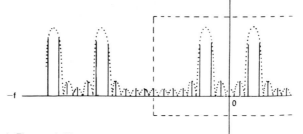

d. The sampled frequency domain and corresponding time domain.

Figure 6–12   (continued)

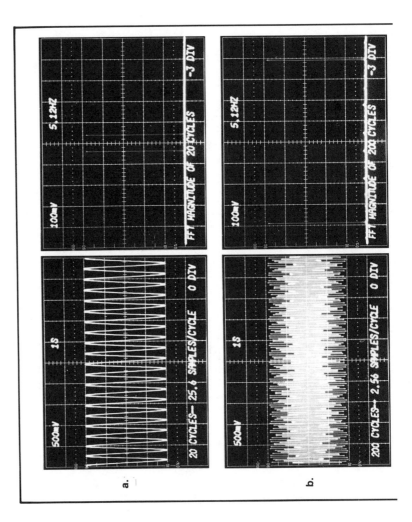

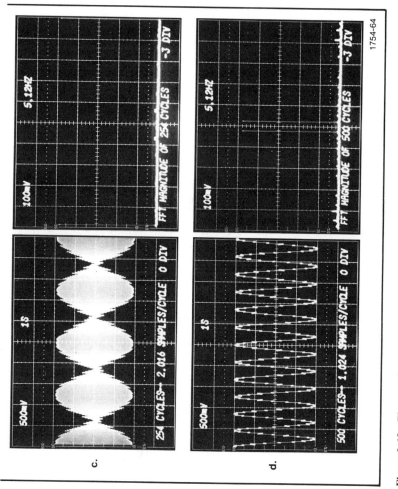

1754-64

Figure 6–13 The sampling rate must provide at least two samples on each cycle of the highest frequency component. Aliasing occurs if there are less than two samples per cycles.

A large jump in frequency from that in Fig. 6–13c puts us well into the region of aliasing. This is shown in Fig. 6–13d, where 500 cycles have been sampled 1.024 times per cycle. Instead of seeing 500 cycles, though, the samples seem to have outlined about 12 cycles of a lower frequency cosine wave. And in the frequency domain, the spectral components have moved back in from the display edges to indicate this low-frequency alias.

Where did the high frequency in Fig. 6–13d go?

Well, it really resides somewhere beyond the edges of the magnitude display.

How did this alias in the lower frequency region come about?

This can be answered by looking at Fig. 6–14.

In Fig. 6–14, 10 cycles of a sinusoid are indicated by a solid line. Let's assume that these 10 cycles represent a high-frequency component, say 100 kHz, of a waveform that is being sampled. Let's further assume that the heavy dots on the 10 cycles represent the amplitude samples relative to the analog-to-digital reference, shown as a solid horizontal line. Notice that there are 12 samples for the 10 cycles of the sinusoid. Since it is a 100-kHz sinusoid and it is sampled 1.2 times per cycle, it follows that the sampling frequency is 120 kHz. We know from recent discussion that this sampling rate is too low for the 100-kHz component. The Nyquist frequency is 120 kHz/2, or 60 kHz, and as a result we should expect to see aliasing of the 100-kHz sinusoid.

The dotted sinusoid in Fig. 6–14 is the low-frequency alias resulting from sampling the 100-kHz component at too low of a rate. Notice how this alias passes through each of the sample points. A little further inspection shows that, for the assumed sampling rate, the dotted sine wave has a frequency of 20 kHz.

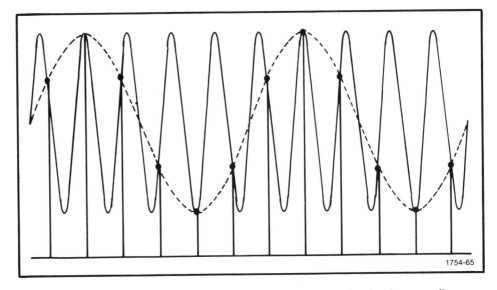

1754-65

Figure 6–14 Insufficient sampling of a high-frequency component results in a low-frequency alias.

Let's look again at the frequencies associated with Fig. 6–14. There is a pattern there associated with aliasing. First, a high-frequency component of 100 kHz was assumed. Then, from the sample arrangement on this 100 kHz component, a sampling rate of 120 kHz was determined. And since the Nyquist frequency is one-half the sampling rate, anything above 60 kHz becomes aliased. *For our example, the Nyquist frequency is 40 kHz below the 100-kHz component.* We also noticed that the sample points describe a low-frequency sinusoid, This low-frequency sinusoid, the alias, has a frequency of 20 kHz—*it occurs 40 kHz below the Nyquist frequency.* Is the pattern beginning to emerge for you?

When a sampled waveform has frequency components above the Nyquist or folding frequency, those components are folded about the Nyquist frequency into the frequency domain below. To aid in visualizing this folding action, refer to Fig. 6–15.

Another demonstration of aliasing is shown in Fig. 6–16. In this case, an ideal square wave was constructed in a waveform array by a software routine. Then the square wave was transformed to the frequency domain by the FFT. The frequency-domain magnitude is shown in Fig. 6–16. Again, the Nyquist frequency is at the left and right sides of the display. The nonaliased components are marked with numbers 1 through 6. All of the other components are aliases of higher frequencies, frequencies that actually exist beyond the edges of the display area. They have been folded about the edges of the display by aliasing and proceed in lower frequencies to 0 Hz (center of the display). At 0 Hz, they fold again and move out to the edges of the display, where they fold again. This goes on as long as there are aliases to be folded.

In general, you can distinguish aliases by knowing something about the waveform you have sampled. If it is periodic, you know that it has harmonic content, and you can find its fundamental from the period. Thus, you can predict where the harmon-

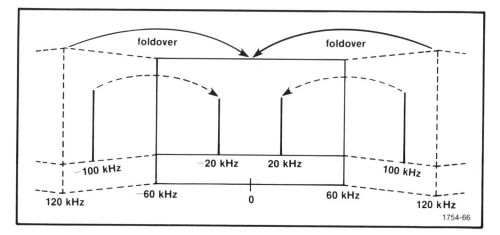

Figure 6–15   When the sampling rate is 60 kHz, a 100-kHz component is folded down to become a 20-kHz alias.

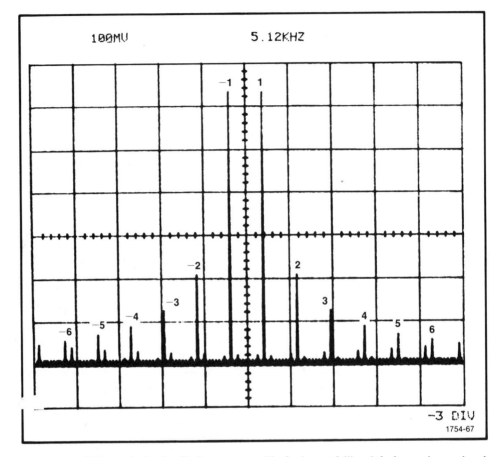

Figure 6–16 FFT magnitude of an ideal square wave. The fundamental (1) and the harmonics numbered 2, 3, 4, 5, and 6 are not aliases. All others are aliases.

ics should be. In general, the aliased components appear between these harmonics.

In the case of nonperiodic waveforms, where the frequency domain appears to duplicate a continuous spectrum instead of a line spectrum, you can be assured that aliasing has occurred if the spectrum hasn't dropped to zero before the frequency display edges. In fact, the higher the amplitude at the display edges, the greater the degree of aliasing. Actually pointing out aliases, however, is difficult, since they fold back and add to the existing low-frequency components.

Aliasing is inevitable when analog waveforms are windowed and sampled. Remember, the data window itself has a spectrum that is infinite in extent. It is impossible to have a sampling rate that assures 2 samples per cycle for all frequencies. In most cases, however, the high-frequency portion of a windowed waveform's spectrum drops to what can be considered an insignificant level at higher frequencies. Then all you need to do is select a sampling rate such that several samples per cycle occur for the highest frequency of significance. Any aliasing that occurs then is below the level of consideration or at least is not resolvable within the limits of computer word length.

There are some cases where waveforms are not high-frequency limited. The ideal square wave whose spectrum is shown in Fig. 6–15 is one example. Waveforms with very fast rise times and responses from high-pass filters are not band limited, and it is difficult to escape aliasing by adjusting sample rate. In these cases, aliasing can be prevented by filtering the waveform before it is sampled. Filters used for doing this are referred to as *antialiasing filters*, and they are designed to limit the high-frequency content of a waveform to a known and acceptable cutoff frequency.

OTHER SOURCES OF ERROR

Noise, leakage, periodicity, and aliasing are all things that are related to implementing the FFT in general. They really aren't FFT errors, though. In truth, they are errors associated with the analog waveform and how it is acquired and sampled.

Besides these general types of errors, there are possibilities for other errors related to the specific hardware and techniques used to acquire data. For example, sampling and digitizing can be done by hand from cathode-ray-tube photos or by using various video techniques. In such approaches, optical errors may become part of the data. The electron optics of a cathode ray tube and the lens optics of camera systems may add geometric distortion to the acquired waveform. Naturally, this distortion becomes part of the data that the FFT works on and may have a significant effect on the frequency domain. There are, however, various software approaches that can provide trace prediction and geometry correction.

Each approach to acquisition and sampling has specific errors unique to that approach. It is impossible to cover all of these approaches and list the possible errors. For the most part, this is not necessary. Most complete and comprehensive measurement systems do recognize these errors and attempt to reduce them to insignificance through proper hardware design and correction algorithms.

Chapter 7 ——————————————

A Guide to Using the FFT

Up to this point a substantial amount of Fourier theory has been covered along with some general principles for applying it. The goal has been understanding. Now these same things are covered again, only this time they are covered in terms of successfully using the FFT. The goal here is improved results from the FFT.

SOME IMPORTANT PROPERTIES OF THE FFT

For the most part, the important properties of the Fourier integral are also the important properties of the FFT. These properties were covered in Chapter 3. However, they are briefly covered again here, but specifically in terms of the FFT. Not only are these properties a key part of understanding FFT results, but they are also a key part of manipulating the FFT for better results.

1. *The FFT has an inverse.* Any sampled waveform that is transformed to the frequency domain by the FFT algorithm can be transformed back to the time domain by basically the same algorithm. This is important for performing time-domain operations that are more easily done in the frequency domain, such as convolution and correlation.

2. *Even functions transform to real parts only.* If a function is windowed so that repeated windows form a function that is a mirror image about time zero, that is, symmetric so that $x(t) = x(-t)$, then the windowed function is an even function. Its frequency domain is real and also even.

3. *Odd functions transform to imaginary parts only.* If a function is windowed so that repeated windows form a function that is an inverted mirror image about

time zero, that is, $x(t) = -x(-t)$, then the windowed function is an odd function. Its frequency domain is imaginary and odd.

4. *Arbitrary functions are the sum of even and odd parts.* Any function may be expressed as being even or odd or the sum of even and odd parts.

5. *The FFT is a linear transform.* This is a property whereby two or more waveforms can be summed in the time domain to give a third function, and the frequency domain of this new function is the sum of the frequency domains of the original functions.

6. *Time scaling affects frequency scaling.* A time-scale expansion corresponds to a frequency-scale compression, and a time-scale compression corresponds to a frequency-scale expansion. The effect of time scaling on the frequency-domain amplitude depends upon the type of waveform acquired (periodic or nonperiodic) and its relation to the window.

7. *Frequency scaling affects time scaling.* A frequency-scale expansion corresponds to a time-scale compression, and a frequency-scale compression corresponds to a time-scale expansion.

8. *Time shifting affects phase only.* Shifting a waveform within the window changes the real and imaginary parts of the frequency domain in such a manner that the square root of the sum of the squares (the magnitude) remains constant. The ratio of the imaginary part to the real part varies, however, and affects phase.

9. *Frequency shifting causes time-domain modulation.* Shifting a frequency-domain function by $+F$ and $-F$ causes a sinusoid of frequency F to be modulated in the time domain by the time-domain function corresponding to the frequency-domain function before shifting.

10. *The convolution property.* Multiplication of two time-domain waveforms corresponds to convolution of each waveform's frequency-domain function. Conversely, forming the complex product of two frequency-domain functions corresponds to convolving the associated time-domain waveforms. Thus, time-domain convolution is quickly performed with the FFT by transforming the waveforms to be convolved to the frequency domain, forming the complex product of these FFT results, and inverse transforming the product back to the time domain.

11. *The correlation property.* Correlation of two time-domain waveforms corresponds to conjugating the frequency-domain function of one of the waveforms and then multiplying this by the frequency-domain function of the other waveform.

12. *The FFT assumes periodicity in all cases.* The FFT assumes that the windowed data repeats with a period equal to the window time. Thus, there are many assumed windows extending to either side of the physical window, and each is an exact duplicate of the physical one.

SOME GUIDELINES FOR IMPROVING FFT RESULTS

By themselves, the properties of the FFT may seem to make a rather dull list. Some or all of them may, at first glance, seem academic and of little use in actually applying the FFT. Experience, however, reveals exactly the opposite to be true. The following guidelines for using the FFT are derived from the FFT's properties. Their usefulness and validity depend directly on the properties of the FFT.

Signal Average to Remove Additive Noise.   Noise by itself is a function. It, too, has a frequency-domain counterpart. And, by the property of linearity, noise that is added to a signal in the time domain is also added to the signal's frequency-domain function.

It is another property of noise that it is generally random and, thus, has an average value that tends to zero in the long term. Because of this, repetitive signals can be acquired a number of times, and each acquisition can be summed and an average formed that reduces the level of additive noise. The improvement of the signal-to-noise ratio gained by this signal averaging is proportional to the square root of the number of waveforms averaged. When the number of averages is expressed as a power of two, this corresponds to a 3-dB improvement for each power of two average. For example, 128 averages can be expressed as 2^7 and corresponds to a $7 \times$ 3-dB = 21-dB improvement, where truly mean-zero noise is involved.

Whenever waveforms can be repetitively acquired, they should be signal averaged. How many times they should be averaged is a question of how noisy they are. As a general rule, you should signal average any repetitive signal at least 32 times. This reduces low-level noise and any noise added by time jitter. If the waveform is moderately noisy, 128 to 512 averages should provide sufficient improvement. If the waveform is practically buried in noise, many more averages may be required for the desired improvement.

It should always be kept in mind, though, that acquiring waveforms and averaging them takes computer time. The more averaging done, the more time taken. So, for some applications, a trade-off may have to be made between the time required for averaging—the number of averages—and the amount of signal-to-noise improvement desired.

Removing the Mean Often Improves Amplitude Resolution.   Many waveforms are such that their mean value is nonzero—they have a DC component. Sometimes waveforms are purposely acquired with a DC bias; at other times, the DC bias is an inadvertent part of the acquisition. The latter case often occurs when ground-reference information isn't properly supplied with the waveform. As a practical example for another situation, a pure sine wave has a zero mean over its period. But, if the sine wave is considered for a time not equal to its period or an integer multiple of it, its mean value is nonzero. Other situations of inadvertent DC bias may occur when the ground reference from a previous acquisition is not cleared by a new ground reference acquisition; the system may then use the old reference, whatever it may be, for the new waveform acquisition.

Whatever the source of the mean value of the waveform samples, this mean is an added component of the waveform. By the linearity property, it is added in both the time domain and the frequency domain. In some cases, it can be so high in amplitude that it overshadows other waveform components in the frequency domain.

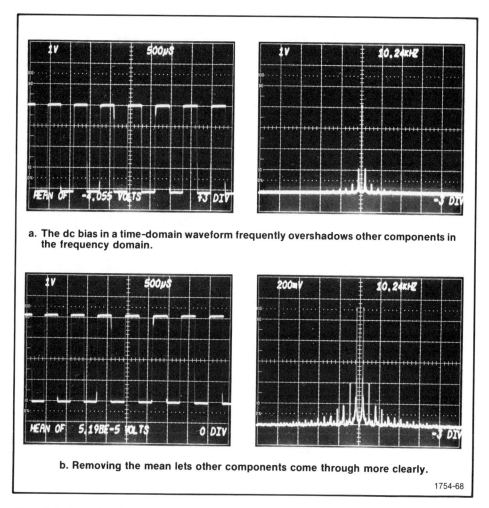

a. The dc bias in a time-domain waveform frequently overshadows other components in the frequency domain.

b. Removing the mean lets other components come through more clearly.

1754-68

Figure 7–1 Removing the mean before transforming a waveform often improves frequency-domain resolution.

Such a case is shown in Fig. 7–1a. To avoid this overshadowing, compute the mean value of the acquired waveform samples. Then subtract the mean value from each sample before doing the FFT. In the case of Fig. 7–1a, removing the mean provides the results shown in Fig. 7–1b.

Always Look at Waveforms from the FFT's Point of View.   No matter what type of signal is acquired, the FFT assumes that the data are repeated at every window length. The effect of this is shown in Fig. 7–2, where signals of several types are first shown in analog form and then with the assumed periodicity of the FFT. The actual segment associated with the window is blocked in solid lines. The assumed repetitions are blocked in dotted lines.

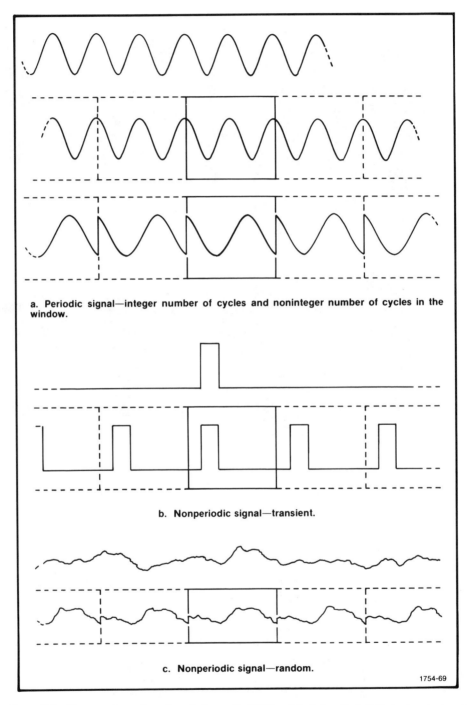

a. Periodic signal—integer number of cycles and noninteger number of cycles in the window.

b. Nonperiodic signal—transient.

c. Nonperiodic signal—random.

1754-69

Figure 7–2   The type of waveform doesn't change the FFT's point of view. Periodicity is always assumed. However, you can change your point of view, according to the type of signal, in interpreting FFT results.

In Fig. 7–2a, the case of periodic signals is shown. Notice that assumed periodicity doesn't change the waveform when the window edges fall on the same relative points of the waveform (integer number of cycles in the window). When the window edges fall at different points on the waveform, however, the assumed periodicity doesn't coincide with the analog waveform's period. This is the case where a noninteger number of cycles appear in the window, and the result is smearing of the frequency-domain information by leakage.

Figure 7–2b shows a nonperiodic waveform of a transient nature. Here, the assumed periodicity causes the transient to appear to be repeated with a period equal to the window length. If we take this periodic point of view, the FFT results describe the discrete frequencies of a repeated pulse's line spectrum. Individual spectral lines may not be apparent, though, since each frequency-domain sample would fall on a spectral line with no interline samples available. If, on the other hand, the nonperiodic point of view is chosen, the FFT results represent estimates of a few discrete samples on the continuous envelope of the energy spectrum for the single pulse. In either case, the FFT results themselves do not change. The numbers are the same. The only thing that changes is the interpretation of these numbers.

Still another type of waveform is shown in Fig. 7–2c. This is a random waveform and is subject to assumed periodicity in the same manner as any other waveform. Here, again, we can take either point of view described for Fig. 7–2b. In both cases, the FFT results for the acquired interval of random data are the same. In no case, however, can we say anything about the data not acquired in the window. We could assume that they are zero outside the window (a pulse), or that the same data exist outside the window as exist inside the window (periodic), or we can make no assumptions at all and just work on the data in the window. Again, these assumptions do not change the results. They just bias our interpretation of them.

Use Sample Rates Greater Than Twice the Highest Frequency. The
Nyquist frequency, F_N, determines the highest frequency component of a waveform that can be defined by sampling. The Nyquist frequency is determined by the sampling rate and is given by $F_N = F_S/2 = 1/2\Delta t$, where F_S is the sampling rate and is equal to the reciprocal of the sample interval, Δt. This works out so that a component at the Nyquist frequency is sampled twice over its period. A component less than the Nyquist frequency is sampled more than twice on each cycle, and one greater than the Nyquist frequency is sampled less than twice per cycle.

Since it takes at least two points per cycle to uniquely define a sinusoid of given amplitude and frequency, any components existing below or at the Nyquist frequency are correctly defined. A component above the Nyquist frequency has less than two samples per cycle and is redefined as a low-frequency alias. For example, sampling a 100-kHz component when the Nyquist frequency is 60 kHz results in the 100-kHz component being aliased. The alias falls below the Nyquist frequency by the amount that the original component exceeds the Nyquist frequency. For the case of the 100-kHz component and the 60-kHz Nyquist frequency, the alias falls 40 kHz below the Nyquist frequency.

In practice, it is impossible to guarantee absolutely stable sampling. So it is not wise to press the two-samples-per-cycle rule too closely. A margin of safety is prudent.

Generally, you should sample so that three or more samples occur for each cycle of the highest expected frequency in a waveform. In other words, as a margin of safety, your sampling rate should be slightly greater than twice the highest significant frequency present in the waveform being sampled.

But you may not always know what the highest frequency in a waveform is, so aliasing may not be consciously avoidable. Or, in some cases, your sampling rate may be constrained by other considerations, and thus aliasing is unavoidable. For these situations, the aliases need to be identified and a choice made between eliminating them from the data or ignoring them. The following are some suggestions for determining if aliasing has occurred and what components are aliases:

1. If the frequency-domain display has significant components near the Nyquist frequency or does not appear to go to zero before the end of the display and remain at zero, aliasing has probably occurred. These conditions are illustrated in Fig. 7–3.

2. If aliases can't be identified and if significant components are near the Nyquist frequency, it is possible that the aliases have folded back on top of and added to valid components. In a line spectrum, this occurs when the edge of the display (Nyquist frequency) falls on a harmonic or midway between harmonics. These possibilities are illustrated in Fig. 7–4.

3. If the frequency domain appears to be a line spectrum and harmonic spacing isn't even or some harmonics seem to increase in magnitude with increasing frequency, aliasing should be suspected. In Fig. 7–5, some uneven spacings

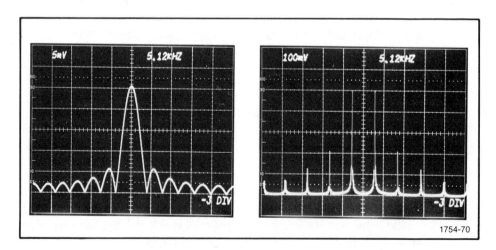

Figure 7–3 Significant components at or near the display edges indicate that aliasing has probably occurred.

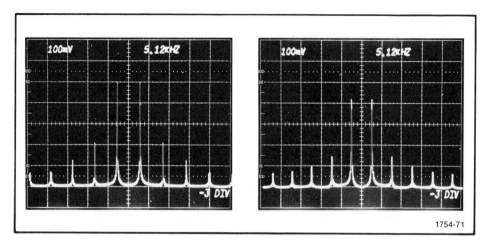

1754-71

Figure 7–4 There are a number of cases where aliases aren't obvious because they are added to valid components.

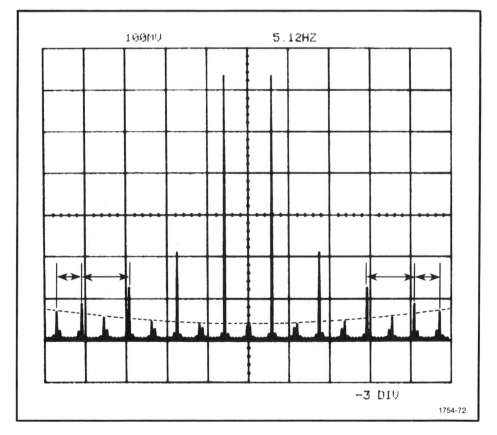

1754-72

Figure 7–5 Component spacing and trends in component amplitudes often aid in identifying aliases.

131

are indicated, and the increasing envelope of aliases is shown by the dashed line. All the components below the dashed line are aliases.

4. In a line spectrum, low-level components residing between high-level components are not always aliases. If these components are located exactly halfway between adjacent, higher-level components, they may be valid harmonics. In the case of a nonsymmetric square wave, the degree of nonsymmetry is indicated by the amplitudes of the even harmonics. This is shown in Fig. 7–6.

Phase Is Only Valid for Existing Components.   Phase is generally computed from the ratio of the imaginary part to the real part of the frequency domain. Quite often, low-level noise between existing components in the real and imaginary parts can give rise to significant phase indications where frequency components don't actually exist. In interpreting computed phase diagrams, remember that phase is only valid at the points where magnitude components are considered to exist.

Remove Delay to Reduce Phase.   From the time-shifting property of the Fourier integral, we know that shifting a waveform relative to time zero affects frequency-domain phase but not magnitude. The amount that a waveform is shifted positively in time is referred to as *delay*. And the more delay there is, the greater the phase.

The same thing holds true for the FFT. However, in terms of the FFT, it is more direct to think of delay as affecting both the real and imaginary parts of the

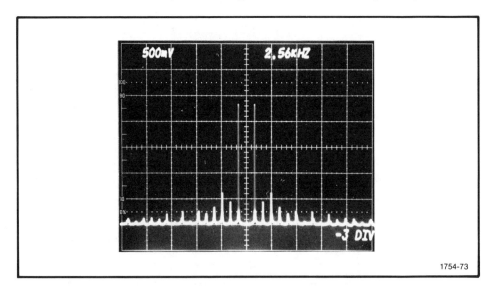

1754-73

Figure 7–6 Don't mistake harmonic distortion for aliases. In the CRT photo above, aliases are at an almost insignificant level. Even harmonics, however, are at very significant levels.

frequency domain. Adding or removing delay causes the real and imaginary parts to change amplitude. Their change relative to each other is such that magnitude ($\sqrt{Im^2 + Re^2}$) remains unchanged. But the ratio of the imaginary to real part does change. Therefore, phase (arctan Im/Re) changes with delay.

By shifting the waveform in the window to reduce delay, the ratio of Im/Re can be reduced. In some cases, symmetries can be taken advantage of to reduce the imaginary part to zero. This corresponds to producing an even function of time and is illustrated in Fig. 7–7.

Figure 7–7a shows a band-limited square pulse that is delayed from time zero. The nonzero real and imaginary parts of its frequency domain are also shown in Fig. 7–7a. The frequency-domain magnitude computed from these is the familiar (sin x)/x magnitude function for a square pulse. Also, since the real and imaginary parts are nonzero, their ratio results in a nonzero phase function.

In Fig. 7–7b, the same pulse has delay removed, so that its midpoint coincides with time zero. Half of the pulse is at the left window edge, and the other half is at the right edge. This is the data arrangement for an even function of time. It is made possible by taking advantage of the assumed periodicity of the FFT, so that repeating the window in negative time completes the left half of the pulse.

As would be expected for an even function of time, the frequency domain in Fig. 7–7b is a real and even function of frequency. The imaginary part, except for digital noise, is zero.

The frequency-domain magnitude computed from the real and imaginary parts in Fig. 7–7b matches that computed for Fig. 7–7a. Magnitude is not changed by time shifting. Phase, however, does change. It is zero (or 180°) if we consider the imaginary part to be absolutely zero. But, since the imaginary part is only effectively zero compared with the real part, we should not expect actual calculations to result in zero phase. The digital noise causes computed phase to be nonzero at many points.

Shifting a waveform in the window to reduce delay can be done during acquisition by choosing an appropriate trigger level. It can also be done after acquisition by using a program to rotate data in the waveform array. In some software packages, a delay argument is provided as an option with the FFT statement. This delay argument lets you specify the amount of delay removal to take place before the FFT algorithm executes.

Change Sample Rates for Different Resolutions.   The rate at which you sample a waveform determines how well that waveform becomes defined in either the time or frequency domain. In general, increasing definition in the time domain causes a decrease in frequency-domain definition. This comes from the reciprocal relationship between the time-domain sample spacing, Δt, and the corresponding frequency-domain sample spacing, Δf. If you decrease Δt for more time resolution, then Δf increases for less frequency resolution, assuming the number of samples in the time window remains the same.

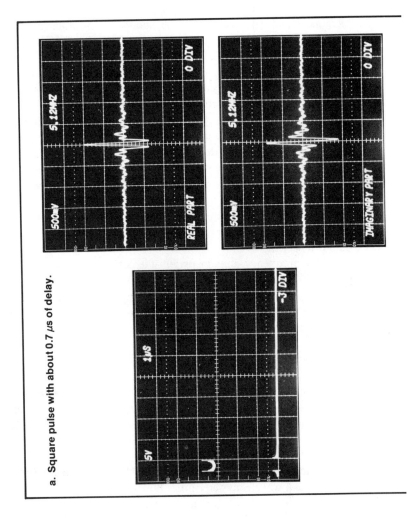

a. Square pulse with about 0.7 μs of delay.

b. The same square pulse with no delay.

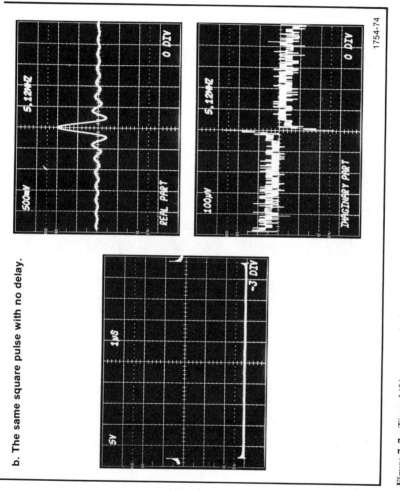
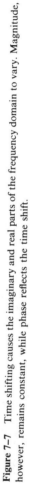

Figure 7–7 Time shifting causes the imaginary and real parts of the frequency domain to vary. Magnitude, however, remains constant, while phase reflects the time shift.

The effect of changing sample rates ($1/\Delta t$ and $1/\Delta f$) is shown in Fig. 7–8, where the number of samples in each display is constant at 512. The sample rates, however, are changed by distributing the 512 samples over different time and frequency intervals.

In Fig. 7–8a, a time-domain waveform is acquired in the window, so that there are many samples over each of its cycles. This is the kind of resolution you would like for studying signals in the time domain. You can see the details of the waveform, and because of a small Δt, you can accurately determine its time-domain parameters. But this also causes the frequency-domain window, Fig. 7–8b, to span more area than needed. As can be seen in Fig. 7–8b, the useful frequency-domain information is concentrated around 0 Hz. The rest of the display contains essentially useless information.

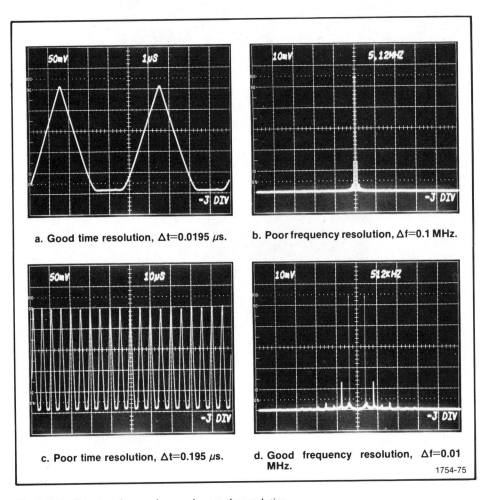

a. Good time resolution, Δt=0.0195 μs.   b. Poor frequency resolution, Δf=0.1 MHz.

c. Poor time resolution, Δt=0.195 μs.   d. Good   frequency   resolution, Δf=0.01 MHz.

1754-75

Figure 7–8   Changing the sample rate changes the resolution.

The opposite situation is shown in Fig. 7–8c and d. In Fig. 7–8c, the same waveform has been acquired so that there are fewer time samples per cycle. This sacrifices time resolution, but look what happens to the frequency domain in Fig. 7–8d. The useful frequency-domain information has spread out to cover most of the display area, and you can easily pick out the different frequency components and their amplitudes. This is what you want for studying signals in the frequency domain.

The choice of sample rate depends upon what you are looking for. If you are after time-domain information only, you'll want to sample for best definition of the waveform, as shown in Fig. 7–8a. If your interest lies in the frequency domain, the approach of Fig. 7–8c is better. However, be careful of aliasing. Generally, your interest will lie in both domains, and a compromise between the two extremes is necessary.

Change Windows to Change Leakage. Leakage is not a universal problem. It does not affect transient data as long as the transient is fully contained in the window. Leakage only occurs when the FFT is used to estimate the discrete line spectra associated with periodic and almost periodic signals. (Almost periodic signals have line spectra, but the components are not harmonically related.) The actual source of leakage, as was described in Chapter 6, is the window used in acquiring the waveform. The amount of leakage depends upon the window shape and how the waveform fits into the window.

For the simplest case, let's consider a periodic waveform in a rectangular acquisition window. If the waveform is acquired so that an integer number of cycles is in the window, leakage won't occur.

It's difficult, however, to obtain exactly an integer number of cycles in a window through standard acquisition procedures. There are just too many variables involved in the process. So leakage is inevitable where periodic waveforms are transformed just as they are acquired. But leakage can be avoided or at least controlled by changing the window to fit the data or to modify the data to a better form.

For example, if about 2.5 cycles of a waveform are acquired, then we can disregard the samples on the offending half cycle and just transform the samples on the desired 2 cycles. In effect, the data window is shortened in the manner of Fig. 7–9a. This illustration shows 2 cycles covered by the first 410 of 512 samples, and leakage can be avoided by doing a 410-point FFT on just the 2 cycles. The usefulness of this technique is, however, limited if the FFT routine is of a type constrained to specific numbers of samples. For example, a power-of-two algorithm can only transform records of 2, 4, 8, 16, . . . , 2^n samples. So a 410-point FFT cannot be done with a power-of-two algorithm.

There is an alternative, though; Fig. 7–9b shows this different approach to shortening the window. Although more complex, it has the advantage of retaining the number of samples needed by a particular FFT algorithm. What is done is the following. The time span of the integer number of cycles is determined first. This is shown as 0.8 sec in Fig. 7–9b. Then you compute the sample spacing needed to

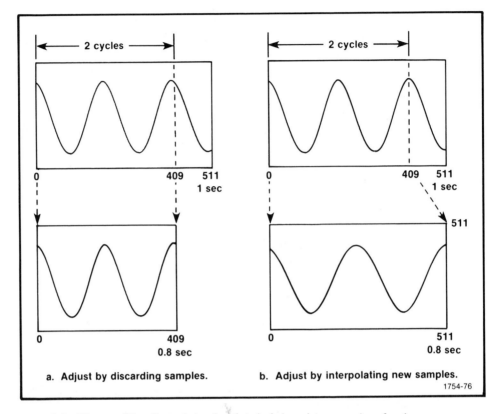

Figure 7–9 When possible, adjust window length to include an integer number of cycles.

place just those cycles in the data window without changing the number of samples needed by the FFT algorithm. For Fig. 7–9b, this is $\Delta t = 0.8/512 = 1.56$ msec. These computed sample locations probably won't match the locations where waveform samples were actually taken. In Fig. 7–9b, the actual samples are located at every $\Delta t = 1/512 = 1.95$ msec compared with the desired 1.56 msec. So the next step is to use the actual sample values and locations to interpolate what the samples values should be for the new sample locations. Linear interpolation is the simplest way to do this, but other methods of interpolation may prove more accurate. The interpolates you compute fill the data window with exactly an integer number of waveform cycles, and leakage is prevented in the transform to the frequency domain.

Since almost periodic data don't have a definable period, the techniques of Fig. 7–9 are not wholly applicable to almost periodic data. They can, however, be applied to make the data at least begin and end at the same level (see Fig. 7–10). This prevents jump discontinuities at the window edges. Eliminating these discontinuities may not entirely eliminate leakage, but it certainly helps to reduce it.

Another way to reduce discontinuities at the window edges, and thus reduce leakage to a tolerable level, is to taper the rectangular window. In short, get rid of

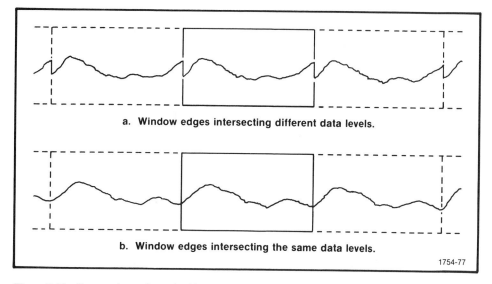

a. Window edges intersecting different data levels.

b. Window edges intersecting the same data levels.

1754-77

Figure 7–10   Remove jump discontinuities by making the data begin and end at the same level.

the abrupt edges; make them fall off smoothly to zero. You can do this by multiplying the acquired data with a window function. Probably the simplest function is the triangular window, and its effect on leakage has already been demonstrated in Fig. 6–9. Some other window shapes that have varying degrees of effect on leakage are shown in Table 7–1.

For the most part, the window shapes in Table 7–1 are explained by their names. The rectangular window, for example, is a square pulse and corresponds to a standard acquisition window. The triangular window is a triangular pulse, and the half sine is the positive going portion of a sine wave. To get the cosine window and extended cosine bell, a $0.5(1 - \cos x)$ function is used in the manner of Fig. 7–11 (p. 142). The remaining windows in Table 7–1 are combinations of other windows. For example, the Hamming window is a 92% cosine added to an 8% pedestal, and the Parzen window is the rescaled convolution of two triangular windows.

To give you a basis for comparing leakage from various windows, Table 7–1 contains the normalized frequency-domain magnitude for each window. Some specific parameters describing these magnitudes are also given. These values were computed from software-generated windows and may vary slightly from theoretical values, but they are very close estimates of theory.

In the fourth column of parameters listed in Table 7–1, the peak magnitude of each window is compared with that of the rectangular window. In the fifth column, the amplitude of the highest side lobe is given in decibels referenced to the major lobe peak. The 3-dB bandwidth of the major lobe is given in the sixth column. These bandwidth values are normalized to β, the reciprocal of the window's time duration. The last column of parameters gives the theoretical rate of decay (roll-off) of the side lobes.

Unity Amplitude Window	Shape Equation	Frequency Domain Magnitude	Major Lobe Height	Highest Side Lobe (dB)	Band-width (3 dB)	Theoretical Roll-Off (dB/Octave)
Rectangle $T=1/\beta$	$A=1$ for $t=0$ to T		T	-13.2	0.86β	6
Extended Cosine Bell	$A=0.5(1-\cos 2\pi 5t/T)$ for $t=0$ to $T/10$ and $t=9T/10$ to T $A=1$ for $t=T/10$ to $9T/10$		0.9 T	-13.5	0.95β	18 (beyond 5β)
Half Cycle Sine	$A=\sin 2\pi 0.5t/T$ for $t=0$ to T		0.64 T	-22.4	1.15β	12
Triangle	$A=2t/T$ for $t=0$ to $T/2$ $A=-2t/T+2$ for $t=T/2$ to T		0.5 T	-26.7	1.27β	12

Window	Formula	Frequency response				
Cosine² (Hanning)	$A = 0.5(1 - \cos 2\pi t/T)$ for $t = 0$ to T		0.5 T	−31.6	1.39β	18
Half Cycle Sine[3]	$A = \sin^3 2\pi 0.5 t/T$ for $t = 0$ to T		0.42 T	−39.5	1.61β	24
Hamming	$A = 0.08 + 0.46(1 - \cos 2\pi t/T))$ for $t = 0$ to T		0.54 T	−41.9	1.26β	6 (beyond 5β)
Cosine[4]	$A = (0.5(1 - \cos 2\pi t/T)^2$ for $t = 0$ to T		0.36 T	−46.9	1.79β	30
Parzen	$A = 1 - 6(2t/T - 1)^2 + 6|2t/T - 1|^3$ for $t = T/4$ to $3T/4$ $A = 2(1 - |2t/T - 1|)^3$ for $t = 0$ to $T/4$ and $t = 3T/4$ to T		0.37 T	−53.2	1.81β	24

Table 7-1 Some common data windows and their frequency-domain parameters.

C1754-79

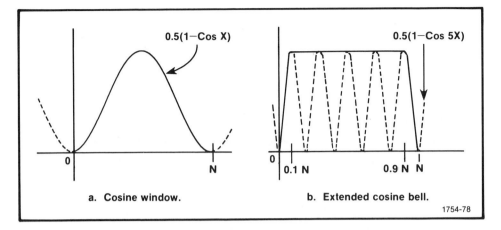

a. Cosine window.   b. Extended cosine bell.

1754-78

Figure 7–11   Generating the cosine and extended cosine bell windows.

    In general, the lower the side lobes, the less leakage, or skirts, in the frequency domain of the windowed data. However, lowering the side lobes also results in more energy being concentrated in widening the major lobe. In Table 7–1, you'll notice that the windows (column 1) are listed in order of decreasing side-lobe level (column 5). As a result, they are also listed in order of increasing bandwidth (column 6). The exception to this is the Hamming window, which has a comparatively narrow major lobe for its side-lobe level.

    In terms of line spectra, the greater the window's bandwidth, the less resolution it provides. In other words, equal-amplitude and adjacent frequencies become more difficult to distinguish. On the other hand, as the side lobes decrease, selectivity increases. This means you have increased ability to distinguish adjacent frequency components of unequal amplitudes. This is further demonstrated in Fig. 7–12 (pp. 144–45).

    Notice in Table 7–1 and Fig. 7–12 that the major lobe magnitude decreases substantially for various window shapes. This is understandable, since each of the unity amplitude windows has less area (energy) than the unity amplitude rectangle. As a result, the frequency-domain magnitude of a windowed waveform decreases according to the window's energy. Using window amplitudes that are greater than unity compensates for this.

    When should windowing be used, and when should it not be used? If windowing is needed, which windowing function should be used? The answers to these questions depend upon what you are looking for. If a waveform has adjacent components of nearly equal amplitude, you may want to leave the data in the rectangular window. The increased major lobe width of another window shape may cause the two adjacent components to leak into each other and appear as one. On the other hand, if there is a small component near a large component, a low side-lobe window will decrease leakage around the large component and make the small component easier to distinguish. Ultimately, selecting a window is a compromise between needed side-lobe reduction and a tolerable increase in major lobe width.

The use of windowing and the choice of windows require some prior knowledge of the signal to be windowed. You have to know what you want to obtain from the frequency domain. And, to a degree, you must know what the frequency domain has to offer. It's much like using light filters in photography. There are some clearcut cases when filters improve the picture. Then there are many other cases that require experimentation before the fine details are arrived at.

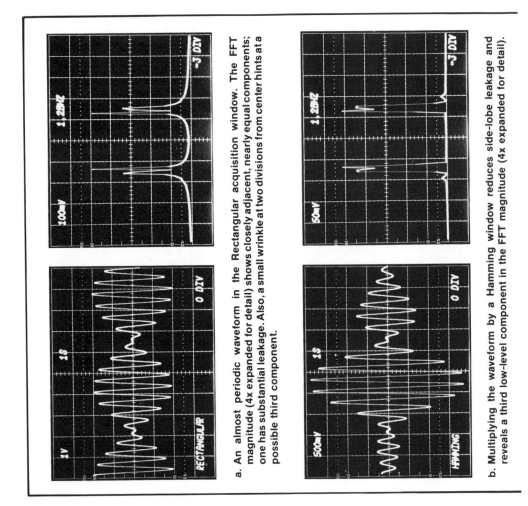

a. An almost periodic waveform in the Rectangular acquisition window. The FFT magnitude (4x expanded for detail) shows closely adjacent, nearly equal components; one has substantial leakage. Also, a small wrinkle at two divisions from center hints at a possible third component.

b. Multiplying the waveform by a Hamming window reduces side-lobe leakage and reveals a third low-level component in the FFT magnitude (4x expanded for detail).

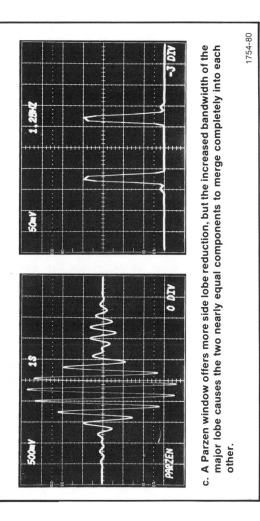

c. A Parzen window offers more side lobe reduction, but the increased bandwidth of the major lobe causes the two nearly equal components to merge completely into each other.

1754-80

Figure 7-12 Windowing is a trade-off between major lobe bandwidth and side-lobe reduction. For this particular waveform, the Hamming window offers the best compromise.

Chapter 8

A Brief Look
at Some FFT Applications

Astronomy, physics, chemistry, statistics, biomedicine, electronics, mechanics, and a host of other related and unrelated fields have areas of study that can and do benefit from the FFT. This is because the FFT is not a discipline-related technique. It is a broad technique of mathematical analysis. Wherever things vibrate, pump, pulse, bubble, burst, or in any other way change with time, there are possible applications for the FFT.

If you are familiar with spectrum analyzers, think of all the places they are used and the literature covering their uses. The FFT is applicable in all of these areas and more. It is only limited by your ability to provide it with the proper data. Given a phenomenon, if you can acquire and sample it, you can FFT it. Or, if you can't acquire it, maybe you can simulate it with a hardware model or a software routine. Whatever the data or its source, the FFT gives you the complex frequency domain, a domain where many difficult time-domain techniques become greatly simplified.

To even briefly discuss all of the application possibilities of the FFT would be a considerable task. In lieu of this, let's glance at just a few representative examples. Perhaps these will suggest additional applications in your specific field of interest.

DISTORTION ANALYSIS

One type of distortion has already been discussed several times. This is the distortion related to square-wave symmetry, where the amplitudes of even harmonics indicate the degree of distortion.

But there are other types of distortion that are probably of more widespread interest. One of these is *harmonic distortion*. Percents of harmonic distortion for various harmonics are often quoted when amplifiers and transmission systems are being discussed. Total harmonic distortion, the sum of the harmonics of distortion, is also often specified.

Testing for harmonic distortion is a relatively straightforward operation. A sinusoid (a pure test tone) is fed into the network to be tested. The frequency of the tone depends on the particular system being tested, with 1000 Hz being a common test frequency for audio circuits. Generally, the level of this tone or sinusoid is set to produce the maximum rated output of the circuit.

The output caused by the test tone contains any harmonic distortion caused by the network under test. If the distortion is appreciable, it can be seen by transforming the output signal to the frequency domain and looking at the frequency-domain magnitude. If the network has caused harmonic distortion, the frequency-domain magnitude will have frequency components that are harmonically related to the test tone. These harmonics are distortion; they are frequencies in the output that are not present in the input. The percent of harmonic distortion can be determined in the manner shown in Fig. 8–1.

Another type of distortion occurs when two test tones are fed simultaneously into a network. Within the network, the signals tend to modulate each other and produce sum and difference frequencies. The production of these sum and difference frequencies (side bands) is called *intermodulation distortion* and is related to amplifier linearity. Determining intermodulation distortion from the frequency-domain magnitude of a network's output is shown in Fig. 8–2.

Of course, any harmonic analysis made via the FFT must be made with due consideration for leakage. You may need to preprocess the data to ensure an integer number of cycles, or you may want to window the data to reduce frequency-domain leakage.

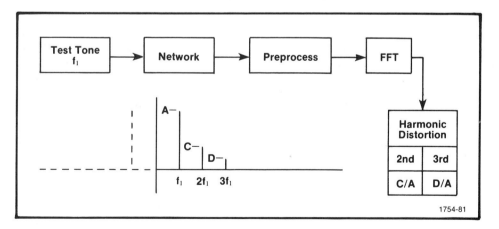

Figure 8–1 Measuring harmonic distortion.

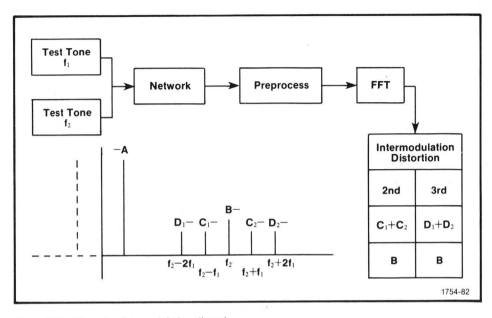

Figure 8–2 Measuring intermodulation distortion.

VIBRATION AND MECHANICAL SIGNATURE ANALYSIS

Motors, pumps, compressors, rollers, and other assorted rotating machinery vibrate when they are in operation. When a machine "vibrates too much," it becomes obvious that "the darn thing is shot." This is the simplest type of vibration analysis. But by the time the results are in, a worn-out $10 bearing may have cost thousands of dollars in damage and unplanned downtime.

A better approach is to monitor machinery vibrations through strategically placed displacement, velocity, and acceleration transducers. Periodic readings of vibration levels can then be taken, and foreboding trends can often be spotted long before the vibrations reach a level that a machine operator would consider suspicious.

Still, vibration level alone cannot tell the whole story. A component can be defective and failing or may have already failed without seriously affecting vibration level. Yet those defect or failure vibrations are there, and they often become quite visible when the machine's total vibration waveform is transformed to the frequency domain. Imbalances, misalignments, and bearing instabilities all add their components to the spectrum.

A vibration spectrum is referred to as a *mechanical signature*, and the determination of information from it is referred to as *signature analysis*. Generally, mechanical signatures are obtained when a machine is placed in operation, or a standard signature is obtained from a known good machine. Then mechanical signatures taken at later dates can be compared with the standard and significant changes in signature compo-

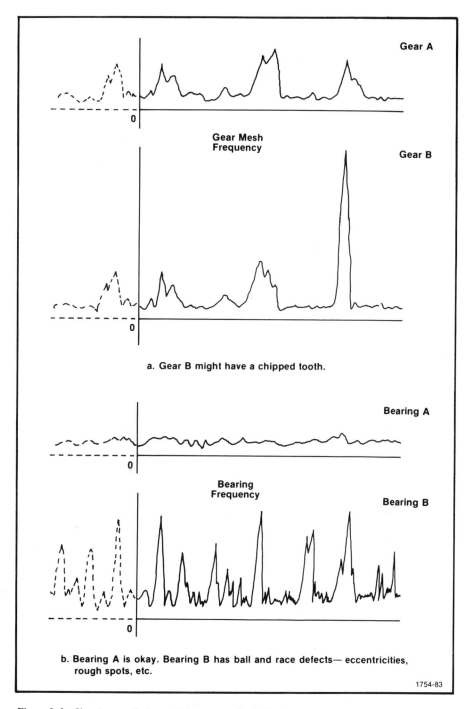

a. Gear B might have a chipped tooth.

b. Bearing A is okay. Bearing B has ball and race defects— eccentricities, rough spots, etc.

1754-83

Figure 8–3 Signature analysis can point out mechanical deficiencies.

nents noted. Often, too, specific signature components can be matched to specific mechanical parts. Thus, not only can defects or impending failure be predicted, but the defective part can also be pinpointed. Some examples of how failures or defects might be spotted are illustrated in Fig. 8–3.

FREQUENCY RESPONSE ESTIMATION (TRANSFER FUNCTIONS)

A large part of scientific and engineering work centers on systems. These systems can be simple or complicated. They can be mechanical, electrical, or biological. But whatever they may be, the main issues are usually "What does it do? How does it react to this stimulus? What would happen if . . . ?"

These issues can often be settled by actually testing the system. Just as often, though, testing the system under the conditions in question may be impractical, or there may be an element of danger involved. Then you would rather predict the test results or at least get some reassuring predictions before actual testing is done.

What you need is a means of completely characterizing the system. You need to know how it will respond to each frequency component of an arbitrary input signal.

How a system reacts at every frequency is called the *system frequency response*. This frequency response is often expressed as an amplitude-frequency plot and a phase-frequency plot or, where possible, a set of equations describing these plots. For a linear, time-invariant system, the frequency response completely characterizes the system. If the system is nonlinear or time varying, a frequency response plot can characterize the system for some specific operating conditions.

There are a variety of methods for obtaining a system's frequency response. The simplest is to feed sinusoids at various frequencies, one at a time, into the system. Then the amplitude and phase changes at the output are plotted. Swept frequency oscillators and various detection schemes are a more common means of obtaining frequency response plots. A spectrum analyzer with a tracking generator is another standard approach. Typically, however, standard instrumentation is limited to providing only an amplitude-frequency plot. Also, swept frequency oscillators and tracking generators have limited frequency ranges. These limitations can result in incomplete characterization of the system—only an amplitude-frequency plot is obtained and only for a limited frequency band. Depending on your analysis job, you may need more.

Another point to consider is that sinusoids are not practical test signals for some situations. This is particularly true for many mechanical and geological systems. Consider oil exploration, for example. How do you apply a sinusoid to the earth's surface? Instead, geological surveys are routinely conducted by exploding small charges on the earth's surface. Responses, picked up by geophones placed at strategic locations, are then used to characterize subsurface structure.

A sharp concentration of energy, such as a geologist's test charge, is generally referred to as an *impulse*. How a system reacts to an impulse is referred to as the *impulse response*. The impulse response is the time-domain counterpart of the frequency response.

If you Fourier transform or FFT a system's impulse response, you get the system's *frequency response*, which is also often referred to as the system's *transfer function*.

In the theoretical sense, an impulse is a function of infinite amplitude, zero width, and unity area. Its frequency domain has unity amplitude at every frequency. In practical terms, however, an impulse has a finite amplitude (great enough to elicit a response but small enough to avoid damaging the system) and a nonzero width. Though nonzero, the width must still be much less than the expected response time of the system.

All of this can be likened to a Chinese gong—the strike of the mallet is the impulse, and the vibration of the gong is the impulse response. If the gong is struck with too much force, it's driven through the wall and destroyed. If it's struck too lightly, the response is minimal and you hear nothing. If you strike it forcefully but let the mallet rest on the gong, it sounds mushy. Only when it is struck sharply and with moderate force does it respond with its clear, characteristic ringing.

There are also cases where an impulse is not an appropriate test signal. For example, in testing electronic networks, it is often easier to apply a unit step function to the input. Then the voltage at the output is referred to as the *step response*. This is related to the impulse response because the derivative of a step is an impulse. And, for a linear, time-invariant system, the derivative of the step response is the impulse response. And, again, frequency response is the FFT of the impulse response. (Note: Most signal processing software packages have differentiation routines that allow you to quickly and easily compute the derivative of any stored waveform.)

There are some cases, too, where none of the standard test signals are appropriate—sinusoids, swept frequencies, impulses, and steps are out of the question. A communications network might be a good example of this.

Suppose the network has already been installed and is in use. You don't want to interrupt it for frequency response testing. So the operating signals must be your test signals. Their relationship to the frequency response is shown in Fig. 8–4, where $x(t)$ is the input signal, $h(t)$ is the impulse response, and $y(t)$ is the output signal.

The simplest relationship between $x(t)$ and $y(t)$ exists when these terms are looked at in the frequency domain. Then the frequency domain of the output, $Y(f)$, is equal to the product of the frequency response, $H(f)$, and the input frequency domain, $X(f)$. The frequency response can be determined by acquiring $x(t)$ and the corresponding $y(t)$ and fast Fourier transforming both to the frequency domain. $H(f)$ is then found by dividing $Y(f)$ by $X(f)$. However, caution is advised in this operation, since there is a potential divide-by-zero situation in computing $Y(f)/X(f)$. A divide-by-zero situation can generally be avoided, though, by placing data checks and branches in your program prior to the point where division takes place.

In fact, a general word of caution regarding frequency response computation and using the other analysis techniques presented in this chapter is probably appropriate here. First of all, it is best to say that the frequency response is being *estimated*. This is because digital techniques are used to represent analog phenomenon. Even if the digital estimate is exact, it is only exact for the sample points. Anything that happens between samples must be speculated from the points on either side. Also, physical systems do change. They change with time—they age—and they usually are only linear in their normal operating regions, and then not always exactly linear. So, even with the best of data acquisition, the most accurate digitizing, and the most precise computations, you should still treat your results as estimates. Maybe Alfred North Whitehead (1861–1947), a prominent mathematician and philosopher of the early 1900s, said it best:

> There is no more common error than to assume that, because prolonged and accurate mathematical calculations have been made, the application of the result to some fact of nature is absolutely certain.

CONVOLUTION

Once you have obtained either the impulse response or the frequency response of a system, you have the system completely characterized. Unfortunately, however, it's usually difficult to tell exactly how a system is going to react to an input waveshape by just looking at the impulse response or the frequency response.

In order to predict the system's output waveshape for a given input waveshape, you need to solve the convolution integral. This is the integral relationship shown in Fig. 8–4. It states that the system's output signal, $y(t)$, is the convolution of the impulse response, $h(t)$, and the input signal, $x(t)$. The τ used in the integral of Fig. 8–4 is just a dummy time variable that facilitates time shifting in the convolution operation. (For a quick review of this time shifting and its relation to convolution, refer to the discussion in Chapter 3 surrounding Fig. 3–16.)

The integral, as shown in Fig. 8–4, can be evaluated in a fairly straightforward

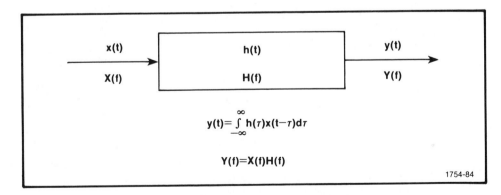

$$y(t)= \int_{-\infty}^{\infty} h(\tau)x(t-\tau)d\tau$$

$$Y(f)=X(f)H(f)$$

1754-84

Figure 8–4 The characterizing parameters of a linear, time-invariant system.

manner in the time domain by using digital techniques. This, however, can be time consuming. It's much quicker and easier to perform convolution by taking advantage of the fact that convolution in the time domain corresponds to multiplication in the frequency domain. The process is to simply transform the input signal to the frequency domain via the FFT to get $X(f)$. Then the impulse response is transformed to the frequency domain to get the frequency response, $H(f)$. The product of these two complex data sets is then formed to get $Y(f)$, which is the frequency-domain function for the system output caused by the input, $x(t)$. The time-domain function for this system output is simply obtained by inverse Fourier transforming $Y(f)$ to get $y(t)$. By knowing the frequency response of a system, the convolution technique described above can be used to predict the system's output waveshape for any input waveshape.

There are some cases, however, when you already know the output waveshape and you'd like to know what input waveshape caused it. This, too, can be estimated as long as you know the $H(f)$ of the system.

The input waveshape can be determined by evaluating $X(f) = Y(f)/H(f)$ for $X(f)$ and then taking the inverse transform of $X(f)$ to get $x(t)$. This operation is often referred to as *deconvolution*, and there is need for caution in using it. Specifically, the complex division of the $Y(f)$ data array by the $H(f)$ data array may result in a divide-by-zero situation at some data points. You'll need to take programming steps to avoid these divide-by-zero possibilities. Also, the deconvolution procedure is sensitive to noise. Noise components in the data being deconvolved may become greatly amplified in the results. However, reasonable improvements can be obtained by applying digital filtering.

Digital filtering is, in the simplest sense, a matter of generating a low-pass filtering function with software. The function is generally constructed as a frequency response. Then the waveform to be filtered is transformed to the frequency domain and complex multiplied with the filter's frequency-domain response. (In the time domain, this corresponds to convolving the filter's impulse response with the waveform.) The frequency-domain product is the frequency domain of the filtered waveform.

The point of filtering is to remove or block unwanted components from a waveform. In the case of preparing waveforms for deconvolution or dressing up the results of deconvolution, the unwanted components are generally those of high-frequency noise. To reduce this noise contribution, a low-pass filter is used. There are a variety of filter functions that can be used, such as elliptical, Chebyshev, Butterworth, and so forth. Choosing and digitally implementing them is another whole topic beyond the basics of the FFT. However, a full discussion of digital filtering is provided in R. W. Hamming's *Digital Filters* (see listing in Bibliography).

As a final note on convolution and deconvolution with the FFT, there are several standard definitions that may be used in the algorithms. These generally differ only in scaling constants, so the results you get from a convolution routine may differ from the expected results by a multiplicative constant. The waveshape will be correct. However, you may have to do some amplitude rescaling.

CORRELATION

These days just about everyone has a feeling for what correlation means. Thanks to news media, politicians, and various special interest groups, correlation has become a household word. We're all probably familiar with headlines and statements similar to the following: "Surgeon General correlates smoking and cancer" and "There is a high degree of correlation between education and income." Thus we've come to equate correlation with terms such as *association*, *cause and effect*, and *similarity*. And, to a large degree, these are good, intuitive definitions for the mathematical operation of correlation.

The mathematical definition of correlation is

$$r(\tau) = \lim_{T \to \infty} \frac{1}{2T} \int_{-T}^{T} x(t)y(t + \tau)dt$$

where $r(\tau) =$ the correlation function formed by summing the lagged products of two waveforms, $x(t)$ and $y(t)$,

$(\tau) =$ the time lag between $x(t)$ and $y(t)$.

Functionally, correlation can be thought of as a matching up of waveform components or a similarity test between waveforms. The equivalent hardware definition in Fig. 8–5 makes this operation easier to visualize.

In terms of digital signal processing, correlation is greatly simplified by using the FFT. The two waveforms to be correlated, $x(t)$ and $y(t)$, are transformed to the frequency domain. Following this, one term is conjugated and then the complex product is formed to give $R(f) = X(f)Y^*(f) = [X^*(f)Y(f)]^*$. Here, an asterisk (*) is used to denote *conjugation*. The final step is inverse transforming $R(f)$ back to the time domain to get $r(\tau)$.

Depending on the waveforms used, two types of correlation can be done. If

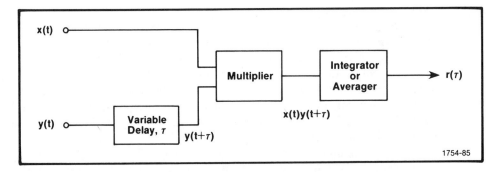

Figure 8–5   Hardware implementation of correlation.

the two waveforms are the same, $x(t) = y(t)$, their correlation is referred to as *autocorrelation*. If the two waveforms are different, $x(t) \neq y(t)$, their correlation is referred to as *cross-correlation*. Let's take a brief look at each of these operations and some possible applications.

Autocorrelation. It is useful to note that the autocorrelation function of a periodic signal is periodic. Also the autocorrelation function of a nonperiodic signal is nonperiodic. These two things are demonstrated in Fig. 8–6a and b, where a sine wave is autocorrelated and random noise is autocorrelated.

The autocorrelation functions in Fig. 8–6 are arranged so that zero time lag is at center screen. Positive time lag is to the right, and negative lag is to the left. In the case of the sine wave's autocorrelation function, maximum correlation occurs at lag zero (center screen). This is where the sine wave is exactly overlaid by itself (a perfect match). Maximum correlation also occurs for the sine wave at every lag equal to the sine wave's period. In Fig. 6–8, however, the autocorrelation function appears triangularly windowed, thereby reducing the maximums at all but lag zero. This apparent triangular windowing occurs because extra arrays of zeros are appended to the waveform arrays before correlation (Fig. 8–7, p. 158). These appended zeros prevent errors from cyclic correlation, but they also make the sine wave appear to be pulsed instead of continuous. This pulsing causes a nearly triangular envelope for the correlation function.

Notice in Fig. 8–6b that the autocorrelation function for noise is large at lag zero and very small for all other lags. An exact match (perfect correlation) is obtained only when the noise exactly overlays itself (lag zero). For other lags, there is little or no match and the noise is said to be *uncorrelated*.

Figure 8–6c shows the autocorrelation function for what appears to be random noise. However, from the periodicity of the autocorrelation function, it is obvious that a periodic signal is buried in the noise. Thus autocorrelation is a useful tool for detecting the presence of periodic signals buried in noise. Biomedical studies, astronomy, and tone-control systems are a few possible application areas for autocorrelation detection techniques.

Cross-Correlation. In autocorrelation, a signal is multiplied by delayed versions of itself. The process of cross-correlation differs only in that two signals are used; one is multiplied by delayed versions of the other. The resulting cross-correlation function contains only those frequency components common to both waveforms.

To see the usefulness of cross-correlation, let's return to the example of detecting a signal buried in noise. Suppose you're receiving signals that are obscured by noise, but you do know the type of signal you are looking for. This is often the case in radar, sonar, and tone control, where the transmitted signal is well defined but the

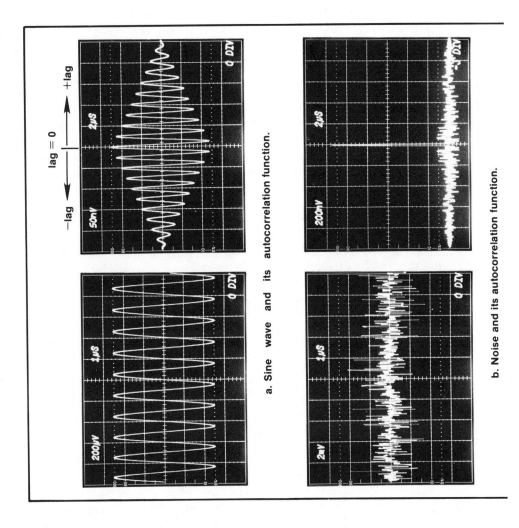

a. Sine wave and its autocorrelation function.

b. Noise and its autocorrelation function.

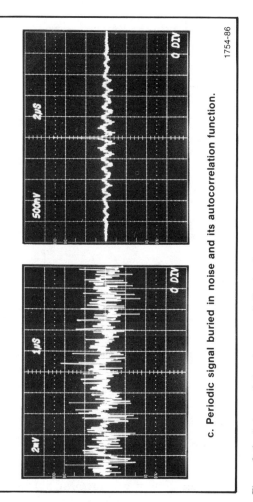

c. Periodic signal buried in noise and its autocorrelation function.

1754-86

Figure 8–6 Autocorrelation detects periodic signals buried in noise.

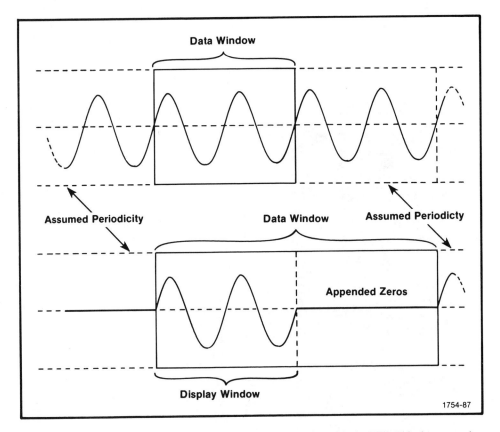

Figure 8–7 Appending an array of zeros prevents cyclic correlation with the FFT. This, however, gives the waveform the appearance of being gated and results in a nearly triangular envelope for periodic correlation functions.

received signal is buried in noise. The situation is demonstrated in Fig. 8–8, where a sine wave buried in noise is illustrated. Notice in Fig. 8–8 that there are no noise components in the cross-correlation function. This is because noise is not common to the signals being correlated.

Speaking of noise, noise is usually a nuisance in most measurement situations. But if you are interested in getting an approximation of a linear system's impulse response, noise can be a useful test signal. All you need to do is drive the system input with a wideband noise source and cross-correlate this input with the resulting system output. If the test is conducted carefully, the cross-correlation results approximate the shape of the system's impulse response.

Besides detecting signals buried in noise and approximating impulse responses,

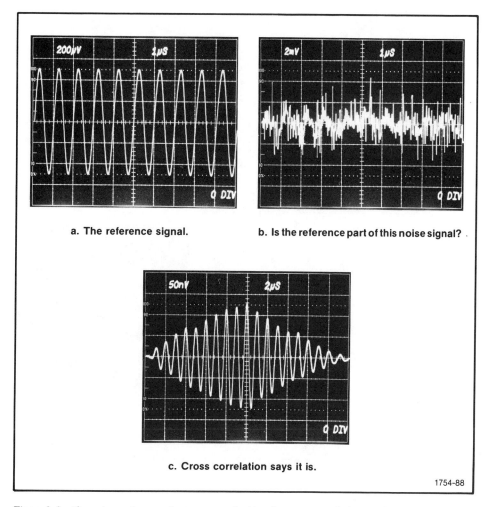

a. The reference signal.

b. Is the reference part of this noise signal?

c. Cross correlation says it is.

1754-88

Figure 8–8 If you know the waveforms you are looking for, cross-correlation can help you find it.

cross-correlation finds many applications where delays must be measured. Time delay is an important parameter in studying path diversity problems, using echo-ranging techniques, or characterizing transmission systems. With cross-correlation, the best match between a transmitted and received signal is found. This best match causes a maximum in the cross-correlation function, and the distance from the maximum to the lag zero point gives the delay between the two signals. The basics of this concept are illustrated further in Fig. 8–9.

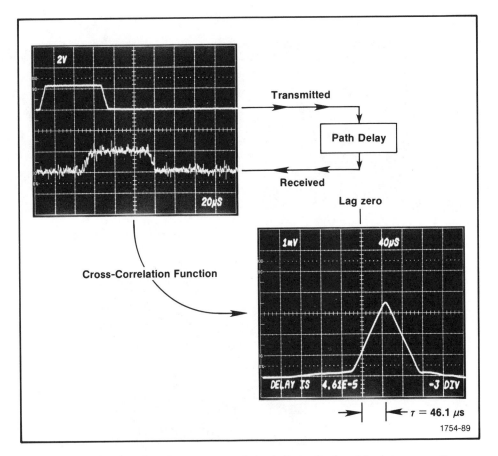

Figure 8–9   The location of maximum cross-correlation indicates the time delay between waveforms.

POWER SPECTRA

As a final note, it should be mentioned that the FFT of a correlation function results in what is called a *power spectrum*. The FFT of an autocorrelation function is generally referred to as the *auto spectrum*. Also referred to as *power spectral density* (PSD), auto spectra are widely used in vibration analysis. When a cross-correlation function is Fourier transformed to the frequency domain, the result is referred to as a *cross-spectrum*. Cross-spectra contain the magnitude products and phase differences of the frequency components common to the signals involved in the cross-correlation. Like auto spectra, cross-spectra are frequently used in vibration analysis too.

Part III

MATHEMATICS OF THE FFT

This part of the book will neither add to nor detract from your ability to use the FFT. So why should you bother with Part III? You ask. Well, first of all, many people have an intrinsic curiosity about what makes things work. For those people, Part III should satisfy some of that curiosity.

Second and more importantly, Part III gives you a basis for communicating with others about the FFT. If you need to discuss various algorithms with a programmer, it will be helpful to be familiar with at least one flow diagram. Also, knowing a little about at least one FFT algorithm is helpful when reading literature on other algorithms and techniques.

Third, as you become more expert in using the FFT, you'll find more people coming to you with questions about how to use the FFT and how the algorithm works. Part III will help you answer some of those questions. And, for the more curious, a Bibliography of more advanced works is provided at the end of this book.

Chapter 9 ————————————————————————

An Algorithm
for Computing the DFT

The FFT is not a single algorithm for computing the discrete Fourier transform. There are several algorithms that provide a basic time advantage over the N^2 operations for the direct evaluation of the DFT. But, because these algorithms are faster than the N^2 approach, they are all lumped under the heading of FFT.

Different FFT algorithms have been developed because various people want to operate on different types of data with different types of machines while exploiting particular properties of the data or machine being used. One particular property of data is the number of samples. Most FFT algorithms are for operating on N samples, where N is equal to 2 raised to an integer power. But some algorithms have been developed to work with N equal to the product of several integers, thus allowing more variation in the values of N that can be handled. For this discussion, however, we'll look at the class of algorithms designed for $N = 2^k$, where k is an integer. The $N = 2^k$ algorithms, also known as *power-of-two algorithms*, are more straightforward and relatively faster to execute than the more general algorithms.

Frequently encountered FFTs are typically based on either the Cooley–Tukey algorithm or the Sande–Tukey algorithm. These two algorithms differ primarily in their organizational approach. The Cooley–Tukey algorithm takes an approach referred to as *decimation in time*, and the Sande–Tukey algorithm uses a *decimation-in-frequency* approach. This latter approach, decimation in frequency, is the approach illustrated here with the power-of-two, Sande–Tukey algorithm for computing the DFT.

THE SANDE–TUKEY ALGORITHM
FOR COMPUTING THE DFT

To begin, let's review the expression for the DFT. This expression was discussed in Chapter 4 and was given there as

$$X_d(k) = \frac{1}{N} \sum_{n=0}^{N-1} x(n) e^{-j2\pi kn/N}$$

For notational convenience, this can be restated as

$$A(n) = \sum_{t=0}^{N-1} x_0(t) W^{-nt}$$

for $n = 0, 1, \ldots, N - 1$. The time-domain data are given by $x_0(0)$, $x_0(1)$, $\ldots, x_0(N-1)$, and W is equal to $e^{j2\pi/N}$. Since the $1/N$ term preceding the summation sign in $x_d(k)$ is simply a scaling term, it is omitted for the sake of simplifying the expressions.

Computing the FFT of $x_0(t)$ consists of $\log_2 N = M$ stages. Each stage requires pairs of computations of the form

$$x_{m+1}(r) = x_m(r) + x_m(s)$$

and

$$x_{m+1}(s) = [x_m(r) - x_m(s)] W^{-p}$$

for specified integers r, s, and p between 0 and $N - 1$ and m between 0 and $M - 1$. The results at the end of the mth computational stage are denoted by $x_m(t)$, where $t = 0, 1, \ldots, N - 1$. Also, the example algorithm is the in-place algorithm. This means that the current results replace the previous results. Thus, $x_{m+1}(t)$ overwrites $x_m(t)$ in going from stage m to stage $m + 1$. The subscript m merely defines a sequence of arrays that define the contents of an associated storage location at the end of the mth stage.

To gain an idea of how the computations take place, consider a 16-point FFT as an example. A superficial look is enough to see the essence of the algorithm. If you are interested in actually implementing an FFT algorithm, several of the references listed in the Bibliography offer more detailed explanations.* The discussion here gives a general idea of the algorithm and some of the terms used in describing FFT algorithms.

The organization of the FFT computations is shown in Fig. 9–1. Notice that the data are gathered into groups at each stage, and, as each stage is passed, the groups are broken into smaller groups. This goes on until the transformation is complete, with one datum per group. This total operation is referred to as *decimation*

---

* E. Oran Brigham's *The Fast Fourier Transform* (see Bibliography) makes an excellent basic reference on FFT algorithms. In fact, Chapter 11 of Brigham's book, and in particular Fig. 11–3, served as the reference for constructing the 16-point FFT algorithm flow diagram graphed in Fig. 9–2.

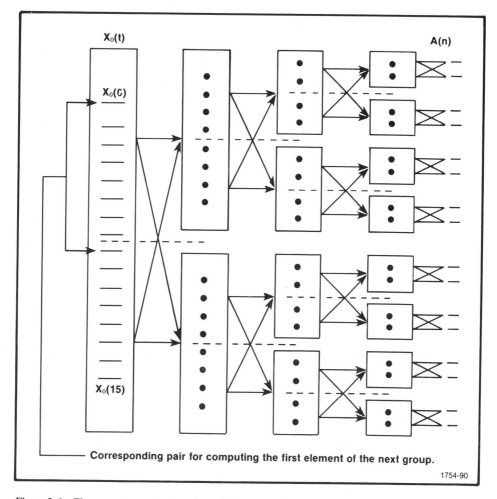

Corresponding pair for computing the first element of the next group.

1754-90

Figure 9–1 The general organization of a decimation-in-frequency algorithm.

in frequency. A decimation-in-time algorithm is organized in exactly the opposite manner—the computations proceed into larger and larger groups.

The data elements within each group of a decimation-in-frequency algorithm are computed from pairs of corresponding elements from the preceding groups. This is indicated in Fig. 9–1 and shown with more detail in the flow diagram of Fig. 9–2. Each of the heavy dots in Fig. 9–2 represents a point of computation. Although these dots are spacially separated in Fig. 9–2, they are one and the same point or memory location in an in-place computation.

As previously mentioned, the computations take place in pairs and are of the form

$$x_{m+1}(r) = x_m(r) + x_m(s)$$

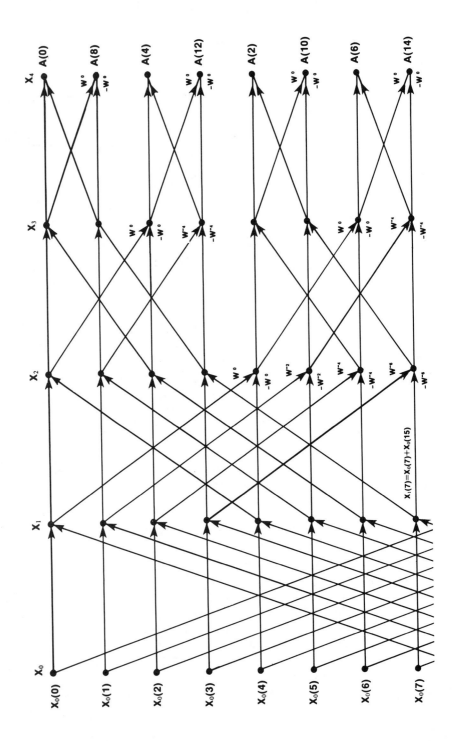

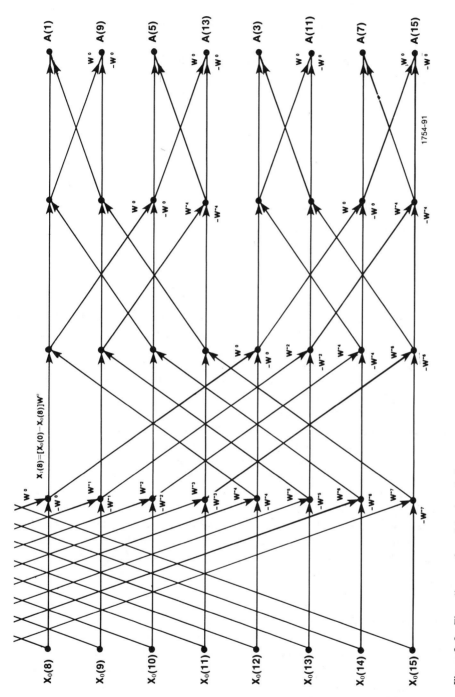

Figure 9–2 Flow diagram for a 16-point, decimation-in-frequency FFT algorithm.

1754-91

A(0)	= A(0000)	bit	A(0000)	= A(0)
A(8)	A(1000)	reversal	A(0001)	A(1)
A(4)	A(0100)		A(0010)	A(2)
A(12)	A(1100)		A(0011)	A(3)
.	.		.	.
.	.		.	.
.	.		.	.
A(11)	A(1011)		A(1101)	A(13)
A(7)	A(0111)		A(1110)	A(14)
A(15)	A(1111)		A(1111)	A(15)

Table 9–1   Bit reversal places the fourier coefficients into correct order.

and

$$x_{m+1}(s) = [x_m(r) - x_m(s)] W^{-p}$$

An example of such a computational pair is shown in Fig. 9–2 at the x_1 stage. There $x_0(7)$ and $x_0(15)$ are added to obtain $x_1(7)$. Also, at the next node, $x_0(0)$ is multiplied by W^0 and added to the product of $x_0(8)$ and $-W^0$ to obtain $x_1(8)$.

Notice in Fig. 9–2 that the operations are grouped at each stage according to whether or not the elements must be multiplied by the W^{-p} factor. This organizational efficiency is further augmented by taking advantage of sine and cosine symmetries to obtain the "twiddle factors," indicated by W^0, W^1, and so forth. Depending on the particular FFT algorithm, these twiddle factors may be individually generated at each stage, or they may be called from a previously generated table.

After the final stage of the algorithm has been passed, $X_0(t)$ will have been transformed to $A(n)$, and the time-domain data will have been transformed to frequency-domain data. As can be seen from the final stage in Fig. 9–2, $A(n)$ occurs in a scrambled order. The order of these Fourier coefficients can be unscrambled by a process referred to as *bit reversal*. If the subscripts of $A(0)$ through $A(15)$ are re-expressed in binary code, the correct location of each frequency component can be determined by reversing or flipping the address bits. This is demonstrated in Table 9–1 and is the final step of the transformation.

A BASIC PROGRAM FOR FFT–IFT COMPUTATION

The preceding steps for FFT computation have been implemented in the BASIC program listed in Fig. 9–3. This program provides a means for additional study and experimentation with the FFT material presented in this book.

It should be pointed out, however, that the FFT program in Fig. 9–3 is intended only as a sample, something for preliminary experimentation. Any serious application

```
100 REM                B:FFTIFT.BAS
101 REM
105 PRINT "RECORD LENGTH MUST BE A POWER OF 2."
110 INPUT "INPUT RECORD LENGTH.";N
115 M=LOG(N)/LOG(2)
120 DIM INPUTRE(N-1),INPUTIM(N-1),BUFFERRE(N-1),BUFFERIM(N-1)
125 DIM TFRE#(N/2-1),TFIM#(N/2-1)
130 REM     ****INPUT DATA FOR XFORM HERE--REAL DATA INTO ARRAY
135 REM      INPUTRE AND IMAGINARY DATA INTO INPUTIM.  IF YOU
140 REM      HAVE REAL TIME-DOMAIN DATA ONLY, SET INPUTIM TO
145 REM      ALL ZEROS.****
150 PRINT "IS THIS AN INVERSE OR FORWARD"
155 PRINT "TRANSFORM? ENTER I OR F."
160 INPUT TYPE$
165 IF TYPE$="I" THEN SIGN=1
170 IF TYPE$="F" THEN SIGN=(-1)
175 IF SIGN=-1 THEN 305
200 REM
201 REM ****REORDER DATA FOR INVERSE FFT****
202 REM
205 FOR I=0 TO N/2
210 BUFFERRE(I)=INPUTRE(I+(N/2-1))
215 BUFFERIM(I)=INPUTIM(I+(N/2-1))
220 NEXT I
225 FOR I=0 TO N/2-2
230 BUFFERRE(I+(N/2+1))=INPUTRE(I)
235 BUFFERIM(I+(N/2+1))=INPUTIM(I)
240 NEXT I
245 FOR I=0 TO N-1
250 INPUTRE(I)=BUFFERRE(I)
255 INPUTIM(I)=BUFFERIM(I)
260 NEXT I
300 REM
305 REM ****GENERATE TWIDDLE FACTORS****
310 REM
315 PI#=3.141592653589795#:PI2#=2*PI#
320 FOR P=0 TO N/2-1
325 TFRE#(P)=COS(PI2#*(-P)/N)
330 TFIM#(P)=(SIGN)*SIN(PI2#*(-P)/N)
335 NEXT P
400 REM
405 REM ****COMPUTE FAST FOURIER TRANSFORM****
410 REM
415 FOR I=1 TO M
420 L=0:H=0
425 G=(N/2^I)
430 FOR K=0 TO (N-1) STEP G
435 TFI=0
440 TFIFLAG=(-1)^(L+1)
445 FOR J=0 TO (G-1)
450 TFI=J*2^(I-1)
455 R=K+J:S=J+H:T=J+G+H
460 IF TFIFLAG>0 THEN 480
465 BUFFERRE(R)=INPUTRE(S)+INPUTRE(T)
470 BUFFERIM(R)=INPUTIM(S)+INPUTIM(T)
475 GOTO 500
480 TEMPRE=INPUTRE(S)-INPUTRE(T)
485 TEMPIM=INPUTIM(S)-INPUTIM(T)
490 BUFFERRE(R)=TEMPRE*TFRE#(TFI)-TEMPIM*TFIM#(TFI)
```

Figure 9–3 MBASIC program for computing the forward and inverse FFT.

```
495 BUFFERIM(R)=TEMPRE*TFIM#(TFI)+TEMPIM*TFRE#(TFI)
500 NEXT J
505 L=L+1:H=INT(L/2)*G*2
510 NEXT K
515 FOR II=0 TO N-1
520 INPUTRE(II)=BUFFERRE(II)
525 INPUTIM(II)=BUFFERIM(II)
530 NEXT II
535 NEXT I
540 FOR I=0 TO N-1
545 INPUTRE(I)=INPUTRE(I)/N
550 INPUTIM(I)=INPUTIM(I)/N
555 NEXT I
600 REM
605 REM ****BIT REVERSAL ROUTINE TO UNSCRAMBLE FFT RESULTS****
610 REM
615 FOR I=0 TO N-1
620 INDEX%=I
625 IOUT%=0
630 FOR J=1 TO M
635 TEMP%=1 AND INDEX%
640 IOUT%=IOUT%*2
645 IOUT%=IOUT%+TEMP%
650 INDEX%=INDEX%\2
655 NEXT J
660 BUFFERRE(I)=INPUTRE(IOUT%)
665 BUFFERIM(I)=INPUTIM(IOUT%)
670 NEXT I
675 IF SIGN=1 THEN 815
700 REM
705 REM ****ORDER FFT OUTPUT FOR NEG. FREQ. AT 0 TO
710 REM N/2-2, DC AT N/2-1, POS. FREQ. AT N/2 TO N-2,
715 REM AND NYQUIST AT N-1.****
720 REM
725 FOR I=0 TO N/2
730 INPUTRE(I+(N/2-1))=BUFFERRE(I)
735 INPUTIM(I+(N/2-1))=BUFFERIM(I)
740 NEXT I
745 FOR I=0 TO N/2-2
750 INPUTRE(I)=BUFFERRE(I+(N/2+1))
755 INPUTIM(I)=BUFFERIM(I+(N/2+1))
760 NEXT I
765 REM
770 REM     ****FFT OUTPUT IS HERE IN ARRAYS INPUTRE FOR REAL
775 REM     PART AND INPUTIM FOR IMAGINARY PART****
780 REM
785 STOP
800 REM
805 REM ****SCALE IFT OUTPUT BY FACTOR OF N***
810 REM
815 FOR I=0 TO N-1
820 INPUTRE(I)=BUFFERRE(I)*N
825 INPUTIM(I)=BUFFERIM(I)*N
830 NEXT I
835 REM
840 REM     ****IFT OUTPUT IS HERE IN ARRAYS INPUTRE FOR REAL
845 REM     PART AND INPUTIM FOR IMAGINARY PART****
850 REM
855 STOP
```

Figure 9–3 (continued)

work with the FFT should be done with an assembly-level routine or hardware-implemented FFT. The primary reasons for this are speed and convenience. The program in Fig. 9–3 is written in interpretive BASIC, which will run much slower than a compiled, assembly-level, or hardware-implemented routine. Also, the program in Fig. 9–3 is not written for execution efficiency. It is written primarily to be easily followed and understood.

For those who choose to enter and experiment with the program in Fig. 9–3, the program is written in Microsoft's (TM) MBASIC, a CP/M version of BASIC–80. Every attempt was made to use only the more standard BASIC commands and structures. This undoubtedly eliminated some features of MBASIC that would have made the program more efficient. However, it also makes the program more directly convertible to other versions of BASIC.

In converting the program to another version of BASIC, there are several potential trouble spots that should be pointed out. The first spots occur at line 125 and lines 315 through 330, where the symbol # appears as the last character in several variables. In MBASIC, the symbol # specifies double-precision variables and operations. Because FFT computations are recursive, double-precision operations may be required, depending on machine word size, to keep cumulative roundoff error from becoming objectionable. Some notation other than # may be necessary for other forms of BASIC.

Another potential trouble spot occurs in the bit-reversal routine at lines 620 through 650. There, the percent symbol, %, is used in MBASIC to specify integer variables and operations. Also, in line 635, AND is used to provide a binary bit-wise "and" operation, and in line 650, a backslash (\) is used to denote integer division. These may have to be changed to meet the syntax requirements of other BASICs.

Finally, there is array indexing. The program in Fig. 9–3 uses the convention that array indexing begins with zero. Thus, a 32-element array has indices running from 0 to 31, or 0 to $N - 1$, where N is the number of elements. If the BASIC in which you plan to implement the program uses indices starting with 1 rather than 0, you'll need to modify the FOR loop indexing scheme throughout.

As far as operation is concerned, the program in Fig. 9–3 is a power-of-two algorithm. This means that the record length (number of array elements) of the input data for transformation must be 2 to an integer power—that is, length equal to 2, 4, 8, 16, 32, 64, 128, and so forth. The prompt at line 105 serves as a reminder of this, and upon running the program you'll be asked to input the record length (line 110) for use in program structuring.

The input data for transformation must be entered in the areas of lines 130 through 145. Typically, this consists of reading the data from a peripheral storage device such as a disk or diskette. Two arrays have been dimensioned in line 120 for this data. They are INPUTRE and INPUTIM for the real part and imaginary part, respectively. Most time-domain data have only a real part, which should be input to INPUTRE. Since there is no imaginary part, INPUTIM should be set to zero. However, in the case of an inverse Fourier transform (IFT), the data are frequency-domain data and generally have both a real and imaginary part. For this latter case, the real part is read into INPUTRE and the imaginary part into INPUTIM.

At lines 150 and 155 the program asks whether you intend to do a forward or inverse transform. Depending on your selection, the program takes several actions. For an inverse transform, the input data are first reordered (lines 201 through 260). For either a forward or reverse transform, twiddle factors must be generated (lines 305 through 335). However, notice that the twiddle factor sign in 330 (SIGN) is positive for an inverse transform and negative for a forward transform.

After twiddle factor generation, the transform is computed in lines 405 through 555. This routine follows the flow diagram of Fig. 9–2 and ultimately overwrites the input data in INPUTRE and INPUTIM with the raw transform results. The calculations are done in M stages and grouped in succeeding stages in descending group sizes—$N/2$, $N/4$, $N/8$, . . . , N/N. The major task is simply generating and keeping track of indices, whereas the actual additions and twiddle factor multiplications of the FFT take place in only a few lines (lines 465 through 495). The trick is simply getting the right things combined with the right things at each stage and group.

Once the FFT routine completes, execution moves to the bit-reversal routine in lines 605 through 670. In the case of the inverse transform, this is the final ordering of output data. All that remains for the inverse, then, is to branch to lines 805 through 830 to scale the results by N, since a $1/N$ factor was used in the general transform routine at lines 545 and 550. In the case of the forward transform, bit reversal is followed by another routine to order the results from negative frequency, to DC, and through positive frequency to the Nyquist frequency. The data ordering throughout is such that the FFT results can be fed back into the routine, without reformatting, to be inverse Fourier transformed.

Admittedly, this discussion of the FFT algorithm has been brief. But then its purpose is only to instill a small measure of familiarity with the algorithm. If you would like to study the Sande–Tukey algorithm, or any other algorithm, in greater depth, the Bibliography contains a number of good sources for algorithm information.

Bibliography

BENDAT, J. S., and A. G. PIERSOL, *Random Data: Analysis and Measurement Procedures*. New York: John Wiley, 1971.

BERGLAND, G. D., "A Guided Tour of the Fast Fourier Transform," *IEEE Spectrum*, July 1969, pp. 41–52.

BLACKMAN, R. B., and J. W. TUKEY, *The Measurement of Power Spectra*, New York: Dover Publications, 1958.

BRACEWELL, R., *The Fourier Transform and Its Applications*. New York: McGraw-Hill, 1969.

BRIGHAM, E. O., *The Fast Fourier Transform*. Englewood Cliffs, N.J.: Prentice-Hall, 1974.

BRIGHAM, E. O., and R. E. MORROW, "The Fast Fourier Transform," *IEEE Spectrum*, December 1967, pp. 63–70.

CHIRLIAN, P. M., *Basic Network Theory*. New York: McGraw-Hill, 1969.

COOLEY, P. M., and J. W. TUKEY, "An Algorithm for the Machine Computation of Complex Fourier Series," *Mathematics of Computation*. vol. 19 (April 1965), 297–301.

GOLD, B., and C. M. RADER, *Digital Processing of Signals*. New York: McGraw-Hill, 1969.

GRIFFITHS, J. W. R., P. L. STOCKLIN, and C. VAN SCHOONEVELD, eds., "Signal Processing," *Proceedings of the NATO Advanced Study Institute of Signal Processing*. New York: Academic Press, 1973.

HAMMING, R. W., *Digital Filters* (2nd ed). Signal Processing Series. Englewood Cliffs, N.J.: Prentice-Hall, 1983.

IEEE Transactions on Audio and Electroacoustics, Special issue on the fast Fourier transform and its application to digital filtering and spectral analysis, vol. AU–15, no. 2, June 1967.

IEEE Transactions on Audio and Electroacoustics, Special issue on the fast Fourier transform, vol. AU–17, no. 2, June 1969.

JENKINS, G. M., and D. G. WATTS, *Spectral Analysis and Its Applications*. San Francisco: Holden-Day, 1968.

LANGE, F. H., *Correlation Techniques*. London, England: Illiffe Books, Ltd., 1967. Published in the U.S. by D. Van Nostrand Company, New York (date).

OPPENHEIM, A. V., ed., *Applications of Digital Signal Processing*. Signal Processing Series. Englewood Cliffs, N.J.: Prentice-Hall, 1978.

———, and R. W. SHAFER, *Digital Signal Processing*. Englewood Cliffs, N.J.: Prentice-Hall, 1975.

OTNES, R. K., and L. ENOCHSON, *Digital Time Series Analysis*. New York: John Wiley, 1972.

PAPOULIS, A., *The Fourier Integral and Its Applications*. New York: McGraw-Hill, 1962.

Proceedings of the IEEE. Special issue on digital signal processing, vol. 63, no. 4, April 1975.

RABINER, L. R., and B. GOLD, *Theory and Application of Digital Signal Processing*. Englewood Cliffs, N.J.: Prentice-Hall, 1975.

RABINER, L. R., AND OTHERS, "Terminology in Digital Signal Processing," *IEEE Transactions on Audio and Electroacoustics*, vol. AU–20 (December 1972), 322–37.

RAMIREZ, R. W., "Fast Fourier Transform Makes Correlation Simpler," *Electronics*, June 26, 1975, pp. 98–103.

———, "The Fast Fourier Transform's Errors Are Predictable, Therefore Manageable," *Electronics*, June 13, 1974, pp. 96–102.

ROBINSON, E. A., and S. TREITEL, *Geophysical Signal Analysis*. Signal Processing Series. Englewood Cliffs, N.J.: Prentice-Hall, 1980.

ROTH, P. R., "Effective Measurements Using Digital Signal Processing," *IEEE Spectrum*, April 1971, pp. 62–70.

SCHAFER, R. W., and L. R. RABINER, "A Digital Signal Processing Approach to Interpolation," *Proceedings of the IEEE*, vol. 61 (June 1973), 692–702.

SCHWARTZ, M., and L. SHAW, *Signal Processing Discrete Spectral Analysis, Detection, and Estimation*. New York: McGraw-Hill, 1975.

TRIBOLET, J. M., *Seismic Applications of Homomorphic Signal Processing*. Signal Processing Series. Englewood Cliffs, N.J.: Prentice-Hall, 1979.

VAN VALKENBURG, M. E., *Network Analysis* (3rd ed.). Englewood Cliffs, N.J.: Prentice-Hall, 1974.

Index